# Neanderthals in the Classroom

T0144496

**Elizabeth Watts**
Research Group for Biology Education
Institute for Zoology and Evolutionary Research
Faculty of Biological Sciences
Friedrich Schiller University
Jena, Germany

## CRC Press
Taylor & Francis Group
Boca Raton   London   New York

CRC Press is an imprint of the
Taylor & Francis Group, an **informa** business

A SCIENCE PUBLISHERS BOOK

Cover image: "The Thinker" at State Museum of Prehistory in Halle, Germany
Credit: State Office for Heritage Management and Archaeology Saxony-Anhalt; Photographer: Juraj Lipták

CRC Press
Taylor & Francis Group
6000 Broken Sound Parkway NW, Suite 300
Boca Raton, FL 33487-2742

First issued in paperback 2021

© 2020 by Taylor & Francis Group, LLC
CRC Press is an imprint of Taylor & Francis Group, an Informa business

No claim to original U.S. Government works

Version Date: 20190819

ISBN-13: 978-1-03-208885-3 (pbk)
ISBN-13: 978-1-138-47637-0 (hbk)

*For Zoey—my thunder & joy*

*Organic life beneath the shoreless waves*
*Was born and nurs'd in ocean's pearly caves;*
*First forms minute, unseen by spheric glass,*
*Move on the mud, or pierce the watery mass;*
*These, as successive generations bloom,*
*New powers acquire and larger limbs assume;*
*Whence countless groups of vegetation spring,*
*And breathing realms of fin and feet and wing.*

*Erasmus Darwin. The Temple of Nature.*
*1802.*

# Foreword

## Much to Like About Neanderthals in the Classroom

*Neanderthals in the Classroom* (hereafter *Neanderthals*) is an important book for science educators, of many types, situated in many parts of the world. Importantly, the topics of creation science and evolution have been of interest to Elizabeth Watts in many facets of her life, giving her the drive to do the research necessary to write a comprehensive book. Her inquiries have yielded a panorama of the battles that have raged in regard to the inclusion of evolution in a science curriculum for all. Watts colorfully illustrates the struggles of those who are not only opposed to a requirement that they must learn about evolution, but also to others learning about evolution and acting in the world in ways that diminish their identities—those who accept literal understandings of creation as it is laid out in the Bible.

Watts uses multiple methods to present the problem of creationism versus evolution in a deep context informed through a variety of lenses. The narratives Watts presents, including interviews with converted creationists, are compelling and bring central facets of the problem to life. These stories connect well with theories grounded in neuroscience that highlight the centrality to learning of creation and retention of negative emotions, such as anger and fear, emanating from knowledge of evolution undermining the identities of those whose beliefs about creation are grounded in literal interpretations of the Bible, reinforced by families and tight-knit fundamentalist communities.

Watts has examined the creationism versus evolution issues historically—and to the extent to which it is relevant, the legal milestones—legislative and prosecutorial. Watts uses lenses that range from nations and states to schools, classrooms, families, and individuals. In so doing she examines concept learning in a context of emotional flux, resistance in the face of identity change, and physiological changes as individual's autonomic systems detect physical and social danger as learners create anger and fear based on their understandings of the consequences of disobeying tenets of the Bible and interpretations of those who teach religion in a fundamentalist way. Watts' approaches provide the groundwork to consider ways in which teachers enact a curriculum to support conceptual change about understanding evolution and how to ameliorate and be more aware of negative emotions, their origins, and consequences.

Readers of *Neanderthals* consider emotions and their manifestation in the body, especially as it relates to social communication (parasympathetic) and fight-flight actions (sympathetic) of learners (Porges 2011). Watts considers interventions that include the use of heuristics to illuminate characteristics of a class in which students learn about evolution, breathing meditation to ameliorate emotions, and strategies to promote mindfulness to ensure that negative emotions do not stick to ongoing conduct—thereby promoting environments that are not conducive to learning.

Watts is clear that science for all includes all students learning about evolution. She provides ample support for readers to enact curricula to promote learning and for teacher educators and policy makers to build an infrastructure of support for teachers, learners, and families and communities that are affected when evolution and beliefs about creation are conflictive. All stakeholders can learn from careful reading of *Neanderthals*.

## Is Creationism versus Evolution just an American Problem?

In 2009, Hisashi Otsuji and colleagues were presenting a paper on undercurrents of Buddhism in contemporary practices of science education in Japan (Otsuji et al. 2009). The occasion was the first biennial conference of the East Asian Science Education Association, held in Taiwan. Within moments of beginning the address, Otsuji was asked a question about the challenges of creationism while seeking to teach evolution. For a moment, Otsuji looked exasperated at the question being posed. He then responded, to the effect, that Shintoism and Buddhism comprised more than 95% of religious followers in Japan, compared to approximately 1% Christians. Otsuji continued, "creationism is not a big problem in Japan, so we will focus on the problems in Japan." As a segue to Otsuji continuing his presentation, someone in the audience interjected, "creationism is an American problem."

The distribution of religious followers in Japan is strikingly different than the United States, and also the world. A 2017 study of the changing global religious landscape, undertaken by the Pew Foundation, shows that Christians have the largest following and will not be surpassed by Muslims until much later in this century. The present world ranking of religious followers is Christians, Muslims, unaffiliated, Hindus, and Buddhists. Accordingly, demographics suggest there is a potential for creationism to be a worldwide problem, especially within some Christian communities. Based on these demographic trends, the creation—evolution dilemma, which is a central focus of *Neanderthals*, has relevance to science educators throughout the world. In addition, Watts has looked in depth at numerous issues that have considerable relevance to educators (e.g., teachers, students, teacher educators, researchers, policymakers) throughout the world. Putting this another way, the case of creation science and teaching evolution can be considered as a context that shines light on myriad problems. Hence, *Neanderthals* is a useful resource for educators writ large. Watts thoughtfully addresses issues that will no doubt catch the attention of readers as they consider challenges they face, wherever and however, they are situated in global education.

# Freedom to Make Sense of Religion and its Texts

The following quote from Osho (also known as Rajneesh), who is a spiritual leader, provides some remarks on Taoism—extreme perspectives on religion and freedom (Osho 2005, pp. 4–5):

> Christianity, Hinduism, Mohammedanism are super-highways: you need not risk anything, you simply follow the crowd, you go with the mob. With Tao you have to go alone, you have to be alone. Tao respects freedom and not conformity. Tao has no tradition. Tao is a rebellion, and the greatest rebellion possible… . Tao says that if you impose a discipline on yourself, you will be a slave. The discipline has to arise out of your awareness, then you will be a master. If you impose order on your life this will be just a pretension: the disorder will remain deep in the very core of your being; the order will be on the surface, at the center there will be disorder. This is not going to help. The real order arises not from the outside, but from the innermost core of your being. Allow disorder, don't repress it. Face it, take the challenge of disorder—and by taking the challenge of disorder and living it, living dangerously, an order arises in your being. That order is out of chaos, not out of any pattern.

To some extent the quote from Osho shows just how difficult it is for the suggestions provided by Watts to actually succeed—that is, for creationists to understand evolution and change their identities and aspects of their lifestyles. In *Neanderthals*, Watts examines two discourse communities that are referred to as fundamentalism and science. These discourses are incommensurable in a context of teaching and learning evolution—which is a foundational construct in studies of the life sciences. Watts zeros in on a number of salient issues pertaining to problems that are often encountered when science students have a background in living as, and learning about, religion in a fundamentalist way that emphasizes literal understandings of the Bible. Importantly, Watts addresses ways in which emotions such as anger and fear can create barriers to learning science associated with evolution. For such students, the tenets of science clash with religious principles from the Bible, reinforced by parents, siblings, peers from school, and learning from active participation in church activities. Accordingly, much of Watts' framing of problems and approaches to teaching and learning, involves learning, the role of emotion, and associated physiological responses to negative emotions such as anger and fear. More is needed than being aware that religious fundamentalists will be antagonistic to learning about evolution and participating in curricular activities that include evolution.

Watts advocates proactive approaches to teaching and learning that highlight heightening awareness to salient characteristics of teaching and learning evolution when creationism is an issue; ameliorating negative emotions using breathing meditation and increasing mindfulness. In so doing, she employs a variety of frameworks associated with social neuroscience, physiology of the brain, conceptual change, and emotion. Simply put, there is ample to engage and occasionally provoke readers who are situated across a spectrum of methodologies and theoretical frameworks. In the following sections of this Foreword, I describe some of the highlights of this broad sweeping book and hone in on several tenets that provoked me to respond, sometimes to provide an alternative pathway, and on other occasions to elaborate so that readers can follow additional pathways when they have mined the nuggets available in *Neanderthals*.

Following this Foreword are a Preface and an Introduction, which, when I read them, assured me I would enjoy reading more than 350 pages of text. I enjoyed the storied nature of these preliminary chapters. In fact, I was motivated to begin my contribution by expanding the Introduction. Had I done so, this would've been an excessively long Foreword. Suffice to say that the Introductory chapter set the stage for a scholarly journey into a vexing problem that most readers will recognize as creationism versus evolution. Perhaps readers from outside the United States will draw some comfort from what they consider is reading about and learning from other people's problems. Not so fast, all readers are bound to learn from contexts that address numerous hot labels that include perspectives that are: historical, cultural, legal, psychological, evolutionary, neurological, educational, and societal.

## Can Science Teachers Separate Science from Religion?

It is all very well for legislators, policy makers, and school level supervisors to insist on science and religion being separate from one another. I suggest this is a reductive perspective that either regards knowing about science and religion as conceptual or values conceptual ways of knowing over other ways of representing what is known and has been learned. The order to separate science from religion likely refers to speaking and writing about science and religion. Having said that, I can easily imagine combining knowledge of science and religion in art, film and photography, poetry and literature. Activities that permit mergers in these ways would contravene laws and regulations followed in many schools and institutions I have experienced in my career. Ironically, my hunch is that participating in activities that juxtapose and integrate knowledge of science and religion will likely lead to deeper understandings of both domains and especially of the virtues of evolution.

Based on my experiences of scientists, who are religious, there can be benefits from appropriately using knowledge from both domains. I illustrate what I mean by this with a short narrative concerning John DeLaeter, one of Australia's foremost physicists.

John taught me undergraduate and graduate level physics. He undertook cutting-edge research on nucleosynthesis, using mass spectrometers and analyses of isotopic abundance in meteorites and, when they became available, moon rocks. As a youth, John attended a parochial high school associated with the Methodist Church and later with the Uniting Church (an amalgamation of Methodist, Presbyterian, and Congregational churches). In his adult life, John served as a lay preacher and in his teaching of physics he seemingly kept well away from religion. I say seemingly because he somehow conveyed that the stars, planets, and other astrophysical objects he studied, were crafted according to a divine plan that was beyond the realms of science and religion to fully understand. He communicated a sense of wonderment and personal connection to what he was teaching and how he was teaching. His physics classes always felt like physics and never were preachy. And yet, part of John being an excellent role model for me, and I assume hundreds of others, was his ever-present spirituality. His lived experiences as a physicist embraced and presented a sense of a spiritual person who did not parse science from religion. Similarly, in my lifeworld, I seek to do the same. Accordingly, as I read Watts' *Neanderthal*, I consciously resisted suggestions to separate science from religion. My preference was to embrace polysemia and regard them as knowledge systems from which we can learn a great deal—they are different lenses through which we can examine and seek to understand life. It does not seem necessary to elevate one knowledge system over the other.

I am uneasy about insisting that creationists have to learn evolution and maybe change their religious based beliefs about creation. I have friends who are fundamentalist in what they know and do as Christians and I do not see any evidence of them speaking about evolution in the way they are in the world. These folks have had successful careers, and know some aspects of science (e.g., medical science) in what appears to be a viable way. I am thinking of two sisters who have had highly successful careers as business managers. Not knowing and believing in evolution does not seem central to them, and they do not seem to be disadvantaged in life by their literal understandings of the Bible and creation.

Why not use science knowledge for science contexts and religion knowledge for other aspects of life? Let me take the example of fundamentalists coming to a science class to learn evolution. I am uncomfortable with the idea that as an educator, I would seek to have them change their religious beliefs to be consistent with the accepted tenets of evolution. To me this seems like scientism. Instead, I would have a goal that they would understand the science without insisting that what they have learned about religion and creation is wrong or primitive in some way. Instead, I would prefer they understand why they accept what they know of creation through their practice of religion. Additionally, they would learn that science discourse has different facts, which they should seek to understand, and gauge their utility within the domain of science. Learners should also be aware of differences in the two knowledge domains and as their understandings grow, about religion, science, and other knowledge systems they can decide whether and to what extent their lifeworlds change. Of course, with expanding understandings, changes in practices and schemas would likely occur. Learners would be aware of some of what they have learned and how they have altered practices and in other cases they would be unaware.

# It is Reductive to Consider What is Learned about Evolution as only, or Primarily, Conceptual

*Neanderthals* addresses learning and difference through lenses such as misconceptions and conceptual change, with which I have never been entirely comfortable.

Back in the late 1980s I was strongly attracted to radical constructivism and the seminal work of Ernst von Glasersfeld. One of the key tenets that was foundational in this framework was that there were no external truths that were fully knowable to humans—instead, knowledge domains were transcendent. Whatever a human knew and could represent would always fall short of what could be known about a given phenomenon. Also, individuals would make sense of experiences in terms of what they knew already and their own ontologies and axiologies. Epistemology is important too. What is learned by an individual might be represented uniquely and would extend far beyond conceptual knowledge. Having said that, what Elizabeth suggests about conceptual change is grounded in decades of research by superb scholars such as David Treagust and Reinders Duit. In *Neanderthals*, there is a great deal of material that can be appropriated by educators to improve the quality of learning about evolution.

## Learning from Difference

*Neanderthals* approaches difference in a direct and systemic manner. However, for more than 20 years I have worked with colleagues on research using cogenerative dialogue (hereafter cogens). Perhaps cogens can be used to augment what Watts suggests in terms of working to learn with others who are very different from one another. During two decades of research and development, cogens have changed to reflect advances associated with researching new theoretical frameworks. I believe the use of cogens is highly applicable to classes that are characterized by differences like those described in *Neanderthals*. Cogen groups usually consist of 4–6 learners who follow a set of guidelines such as: one person speaks at a time and all others listen attentively; the next person to speak addresses the topic being discussed; new topics are not introduced until there is group consensus that it is time to move on—at least for the moment; each person in a cogen group has approximately the same number of turns of talk; all participants speak roughly the same number of words; all interactions are respectful, and when a transgression occurs it is dealt with by the group. The purposes of cogen are to learn with and from others who are different from one another. A productive way to think about learning in cogens involves four authenticity criteria, which I discuss below.

First, each person should learn something new that reflects a change in ontology. This may be evident in terms of what a person considers to be viable or true. The changes may be represented conceptually, as stories, poetically, or through physical actions. A second criterion is that each person would learn what others know and what they can do. It is not just being aware of what they know and can do, but also, why they consider their knowledge to be viable and its affordances and limitations. The purpose is not to educate others to change to be like one another, but to understand about their knowledge in terms of its importance and limitations. A third criterion is that what is learned, collectively, should be catalytic in the sense that others use what is learned to improve practices. This can include participants in the cogen and those with whom they interact (i.e., ripple effects). Finally, the fourth criterion is that all participants should benefit from their learning equitably. That is, participants in cogen should help one another to benefit/improve.

Throughout *Neanderthals* learning has been broadly considered in ways that embrace emotions. Basically, when a body is physically and socially safe, it is possible for parasympathetic functioning to occur. Parasympathetic functioning is optimal for learning because in this state, the tools for social communication are available—e.g., facial expression of emotion, prosodic features of the human voice, and capacity to differentiate the human voice from other noises (Porges 2011).

## Ameliorating Negative Emotions and Switching the Body to Social Communication Functioning

Watts lists breathing as an ideal way to alter body physiology, so that its monitoring system will register safe physical and social conditions. Our ongoing research provides support for breathing meditation acting as a switch from sympathetic mode to parasympathetic mode, which supports social interaction. That is, a few minutes of breathing can activate a switch from fight-flight mode to social communication mode. Interestingly, the effect can be significantly enhanced if both the inhale and exhale are nasal. To increase the effects of breathing meditation—keep your mouth shut (McKeown 2016).

How does this work? In 1998 Robert Furchgott, Louis Ignarro, and Ferid Murad received the Nobel Prize for medicine, for their research on the role of Nitric Oxide (NO) in the cardiovascular system. Research on the role of NO in the body revealed that approximately 90% of the body's nitric oxide is produced in the nose. If NO is distributed through the airways into the lungs, it expands the blood vessels in the alveoli. This expansion supports more oxygen being carried throughout the body to support respiration. Accordingly, if breathing is nasal, the airways can carry NO to the lungs. If breathing is through the mouth the NO supply is bypassed and the beneficial effects of NO are not experienced.

NO has a short half-life and it can last longer in the body when it is in low concentrations and resides in protected sites, such as the folds in nasal cavities. Research has shown that humming on the exhale increases the concentration of NO in the airways by a factor of 15. Hence, it is desirable to inhale and exhale through the nose and hum on the outbreath. By increasing concentration of oxygen in the blood, individuals can enhance parasympathetic functioning and also improve their health. Numerous studies detail benefits of nasal breathing (inspiration and expiration) and describe deleterious health hazards of mouth breathing, which tends to be habitual (e.g., Lundberg and Weitzberg 1999, McKeown 2016, Severinsen 2018).

Consideration of the benefits of breathing meditation make it clear that its application in classrooms would be potentially useful in breaking the habit of mouth breathing and developing a tool that is helpful in everyday life–especially when a person experiences a build-up of anger or fear, or more generally when a person becomes aware of an imbalance in fight-flight and social communication—tilted toward fight-flight. In more dire circumstances, a person might use breathing meditation to address symptoms such as light headedness, giddiness, and fainting. My intention here is to support Watts in asserting that use of breathing meditation can help creationists learn evolution and more broadly, can ameliorate negative emotions on the fear-anger spectra, if and when they occur. Importantly the meditation tool will also support numerous issues that arise in social life throughout the entire life cycle.

## Active Meditation

In our research, we have also investigated how free writing—or mindfully writing—can be used to free up the mind to facilitate a condition of flow in which persons are able to write fluently on topics of their own choosing. This activity might allow students to freely address thorny issues by writing "what's on their minds." Writing down their perspectives is in some ways an extension of heuristics to heighten awareness about characteristics of an issue that has importance to some aspect of social life—such as speaking and listening. In this case, a 10-minute free write might be scheduled at the beginning of each lesson on evolution. If a person were free to write about whatever they wished, there is a potential that in so doing they will release strong negative emotions and heighten personal awareness about one or more thorny issues. It goes without saying that for an activity like this to have maximal success, students must feel safe that others will not read what they have written unless an author of a free write requests others to read what has been written.

Other active forms of mindfulness have also been used in our work, recently inspired by the writing of Osho (2004). When I attended an 8-day meditation retreat at Osho Tapoban, in Nepal, I was not prepared for what was to come. Even though I had been shown two active meditations by a person who was an experienced Osho follower, I still viewed meditation as a way to still the mind. Instead the active meditation, undertaken with more than 300 others, was a one-hour a day experience that was in many ways shocking. Osho emphasized that meditation involved interactions between the poles of life, moving energy between opposite poles. Within the one-hour meditation there are five phases that are distinctive in their difference from one another. In the first 10-minute phase participants practice chaotically breathing through the nose. Osho (p. 78) notes to "let breathing be intense, deep, fast, without rhythm, with no pattern," always concentrating on the exhalation. The actions are accompanied by music that sets a frenzied pace. A second 10-minute phase, called explosion, is a decided step up from highly active to unbelievably active. Osho exhorts participants to give "your body freedom to express whatever is there…. Go totally mad. Scream, shout, cry, jump, kick, shake, dance, sing, laugh… hold nothing back; keep your whole body moving" (p. 79). Then, in phase three, also for 10 minutes, participants raise their arms high above their heads and jump up and down, landing on the flats of the feet, all the time shouting the mantra "hoo, hoo, hoo." Osho says, "exhaust yourself completely" (p. 79). In the fourth phase, lasting

15 minutes, the music stops and so do participants, who are asked to "freeze in whatever place they are when the music stops. No movement, no sound, and witness everything that is happening to you." Finally, the last 15-minute phase is referred to as celebrate, with music and dance.

I describe the phases of Osho's *Dynamic* because what happens is so counter to typical images of meditation of a guru on an isolated mountain top—maintaining stillness of body and mind. In his extensive writing on meditation and mindfulness, Osho is revolutionary, and above all, he shows that meditation and mindfulness need not be separated from life. As part of life, they help participants to better cope by becoming aware of and breaking down routines. In essence, the end goals relate to freedom/liberation in its myriad forms.

In extending the discussion of meditation and mindfulness my intention is for readers to decide whether any of the active meditation forms have a place in educating citizens to cope with life, and in this context, whether active meditation can assist learners to learn about topics like evolution when pain and suffering are ever-present due to conflicting claims for truth often carrying profound implications for identity and lifestyles. Is active meditation an essential life skill? Or is it a tool for religious fringe-dwellers, such as those who follow the practices and philosophy of Osho?

## Freedom in the Classroom

I have been involved in science education research for almost 50 years. Sadly, the classes I visited in the early 1970s are much the same as those I visited in 2019. For the most part the students are controlled by what the teacher has planned for the lesson. Given the problems we have observed in classrooms, there is usually an excess of emotions, often fear and/or anger, among students and the teacher. In our research, we have seen how excesses of emotion can catalyze dysfunctional learning environments and sustain the dysfunction.

When we first began to use finger pulse oximeters to measure pulse rate and blood oxygenation, we realized that teachers and some students were typically acting in fight-flight mode, and in some circumstances teachers entered the dangerous freeze mode. It was clear that time-outs were in order, to enable teachers to calm themselves—stepping away from the fray of teaching and using breathing meditation to restore parasympathetic nerve functioning.

Since we began these studies we have learned that emotions can be ameliorated not only by breathing meditation but also by lightly holding the fingers and, in prayer fashion, pushing together the palms of the hands. As more research is conducted we are identifying other ways to meditate, using dance and music for example, and to restore harmony in a physiological and social sense using light touch on different parts of the body (Tobin 2018). We are expanding the ways in which participants in educational settings can enhance mindfulness and protect individuals from physical and psychological harm from operating too long in fight-flight and freeze modes.

Why should freedom to meditate if and as necessary only be a right for teachers? Given the nature of learning environments and the stressors in everyday life, it is arguable that one of the rights of learners is to take time out to meditate and engage in mindful practices if and as necessary. Students and teachers should have the right to participate in contemplative activities and continuously monitor their emotional wellbeing and physiological and psychological signs of potential disharmony. Freedom to learn seems like a no brainer and a corollary is that freedom to maintain a healthy mind and body is prerequisite to effectively learning—whether it concerns evolution or other knowledge that is of value.

# References

Lundberg, J.O.N. and Weitzberg, E. 1999. Nasal nitric oxide in man. Thorax 54: 947–952. doi: 10.1136/thx.54.10.947.

McKeown, P. 2016. The oxygen advantage: Simple, scientifically proven breathing techniques to help you become healthier, slimmer, faster, and fitter. New York: William Morrow Paperbacks (an imprint of HarperCollins).

Osho. 2004. Meditation: The first and last freedom. A practical guide to meditation. Kathmandu, Nepal: Osho Tapoban.

Osho. 2005. Never born never died. New Delhi: Full Circle.

Otsuji, H., Tasaki, Y., Taylor, P. and Settelmaier, E. 2009. Undercurrents of Buddhism in contemporary practices of science education in Japan 1: Auto/Ethnography as method. Paper presented at the first biennial conference of the East Asian Science Education Association.

Porges, S.W. 2011. The polyvagal theory: neurophysiological foundations of emotions, attachment, communication, and self-regulation. New York, NY: W. W. Norton.

Severinsen, S. 2018. Nitric oxide—a pleasant poison! Accessed on May 11, 2019 at https://www.breatheology.com/2014/04/01/nitrogen-oxide-pleasant-poison/.

Tobin, K. 2018. The role of mindfulness in harmonizing sustainable lifestyles. Learning: Research and Practice, 4: 112–125. doi: 10.1080/23735082.2018.1435039.

**Kenneth Tobin**
The Graduate Center of the City University of New York

# Preface

I had arguably one of the best biology teachers a kid could have. Growing up in the second poorest school district, in the second poorest city in the United States is not the place one normally expects to find high quality teachers, but Mrs. Berry had chosen to teach at this inner-city school. While gang tensions dominated the other classrooms on the fourth floor, Mrs. Berry's biology classroom was a place where intellectual creativity was fostered. She did not rely on the district or school to provide her with materials but instead went out of her way to create materials and an environment that increased the students' interest in science.

She created comic strips to teach us the organelles, with 'Rip the Ribosome' cast as the Austrian body builder and 'Goldi the Golgi Apparatus', who was a portly Danish girl who could not stop stuffing her face with cake. Although these characters might now be considered politically incorrect, they assisted us in learning and remembering the names and functions of the organelles, now even twenty years later. She also arranged for body builders to come flex their muscles for us during our gross anatomy lessons and for the local pound to embalm cats so that we could practice dissecting. Due to her personal investment in time and energy, we received a more in-depth biology education than some college biology majors.

Yet, while I marveled the amount of passion and energy she invested in teaching (whereas other teachers were simply trying to get their students under control), she was not as popular with all of her students. In fact, her love for science seemed to rub one particular student the wrong way. Sitting next to me in the same biology class was Jim (not his real name). Although he had also enjoyed dissecting the cats, he began to continually doubt her abilities as a teacher and openly argued with her during any lesson related to evolution.

Jim and I were good friends, teammates on the track team and part of the same social circle. Yet despite our common interests, I did not share or understand his issues with evolution. I knew he was a devote Christian, who often toted the popular WWJD (What Would Jesus Do?) bracelets, but weren't all the students at our school Christian? El Paso is a predominantly Hispanic town with a major Catholic vibe, where even the toughest gang members will remove their hats when cruising past the local cathedral. Moreover, we had all grown up in the El Paso area across from the mountains in Ciudad Juárez where we could all clearly read 'LA BIBLIA ES LA VERDAD. LEELA' [The Bible is the Truth. Read It]. So why did Jim perceive an

**Figure 1:**  View from El Paso Texas to Juárez, Mexico where "La Biblia es la Verdad. Leela" [The Bible is the Truth. Read it.] is emblazoned on the mountain. Photo credit: wikimedia commons—panoramio.

insurmountable conflict between his religious views and science, while the rest of us did not?

None of this seemed very pressing at the time, but as the years have passed and my interest in science education has grown, these memories keep returning, causing me to wonder in disbelief how such an remarkably dedicated biology teacher, who had created and collected amazing tools and instruments to teach us about biology, was completely unable to get through to Jim. This personal experience has made it clear to me that science education and acquisition of scientific literacy are not ideas that can be relegated to the classroom. In fact, the preconceptions regarding science that a student brings to a classroom are not only important—but as in cases such as Jim's—they may overshadow even the most arduous efforts made by talented teachers.

These insights acted as an impulse for my later research on creationism, which culminated in my doctoral thesis on the topic many years later. During the defense of my doctoral thesis, one of the professors on my panel asked me, "So what should a teacher do when they are confronted by a creationist student?" I had conducted years of research on the effect of creationism on science education and had read all the arguments against creationist claims—but I did not have an answer to what should or could be done at a classroom level, especially not at an interpersonal level. With my heart pounding, I was able to answer the question to his satisfaction by pointing out that a paper by Kevin Padian in 2010 had shown that the use of cladograms and examples of megaevolution can be very effective in better explaining evolution to students and thus theoretically support students in their acceptance of evolution. While the professor on my panel was happy with my answer, I was not. I had sat in classrooms with creationists and I knew that a better explanation of evolution alone would not suffice in addressing the self-perpetuated misconceptions that these students had brought with them to the classroom.

I knew that in order to address these misconceptions, we would have to first acknowledge that lessons on evolution are not purely intellectual exercises but are in fact highly emotional situations for creationists. This emotional aspect is often forgotten or neglected when talking about scientific literacy as scientific pursuits are often centered on the acquisition of knowledge and cognitive comprehension. While emotions should not play a role in scientific research, it is not as simple in education, where emotions play a large role in a student's ability to learn—meaning that we cannot continue to assert that better illustrations or better examples alone will suffice in addressing or mitigating students' rejection of evolution. The intense emotional states that often arise when creationists are confronted by evolution must first be addressed before real solutions can be found. This book can thus be understood as the much longer answer to question I received during my doctoral defense and my humble attempt to bridge the ever-expanding gap between scientists and the general public in order to create space both in the classroom and society where a true dialogue can take place.

# Table of Contents

# Introduction

This book discusses evolution at the intersection of where fundamentalism meets science. While many people of faith are able to harmonize their religious beliefs and scientific data on evolution, other groups, such as young-earth creationists, maintain a fundamentalist (literal) understanding of the Bible and passionately reject evolution. The creationist phenomenon is most prevalent in the United States, where Christian fundamentalism continues to grow in popularity and there is an active movement focused on undermining the teaching of evolution. The effects of this widespread fundamentalist activity and anti-evolution propaganda can be seen at the personal level and societal level. At the individual level, we see the presence of deep-seeded misconceptions regarding evolution, the nature of science and the role of science in understanding the natural world. At a societal level we see extensive denial of evolution and organized activity aimed at altering or preventing the teaching of evolution. The result is that we have a nation that is simultaneously one of the most scientifically advanced in the world, and a nation whose population has one of the lowest levels of acceptance of evolution among all other developed countries (Miller et al. 2006). The rejection of this central tenet of the biosciences within such a modern and educated society is not only ironic but also troubling. The overt anti-evolution movement in the United States causes confusion among the general American public not only regarding the validity of the theory of evolution, but also with regard to the true nature of science, and thus represents a grave threat to general scientific literacy.

Through the course of this book we will look at the complex mix of societal, cultural and psychological factors that lead to the maintenance of creationist belief systems and then discuss concrete educational and communication tools that can help amend these misconceptions by assisting individuals in the integration of scientific findings about evolution into their worldview without feeling as if they need to choose between their religious beliefs and science. While American biologist Jerry Coyne would frown upon this as an 'accommodationist' strategy, we will see that there are numerous ethical, legal, psychological and educational reasons why we should not want to create an 'either or' situation for creationist students when teaching about evolution.

To begin our analysis, the first half of the book, *Fight or Flight*, is devoted to understanding the complexity of the creationism | evolution conflict. In order to understand the true depth and complexity of this conflict, it is necessary to understand how and why it arose. For that purpose, we will look at the rise of creationism as a product of a complex mix of historical, cultural and societal factors. In the first chapter, *Religion Meets Science*, we take a detailed look at how the creationist phenomenon arose and how it has been perpetuated for over one hundred years. In this chapter, the difference between fundamentalism and religion is defined in order to elucidate how evangelicalism, creationism and fundamentalism differ from mainline Christianity and why identifying this differentiation is a vital when discussing potential educational interventions. We then examine creationism from a socio-historical and cultural point of view in order to understand how the origin of creationism in the United States is related to the general development of religiosity and evangelicalism within the context of American history. Here we look at the United States' unique religious history, which set the proverbial ball rolling towards the U.S. becoming the one of the most religious developed countries in the western world and one that harbors a complex strain of Christian fundamentalism and an active anti-evolution movement. While Theodosius Dobzhansky warned against this dangerous anti-evolution trend among American evangelicals more than forty years ago, the complex resistance to evolution has not diminished over the past four decades. Instead, evangelical beliefs only appear to have increased in popularity and variety during that time. It is therefore necessary to look at how the rise of evangelicalism has led to organized fundamentalist and political activities that have had direct and indirect effects on the nation's educational system and the public perception of science.

In the subsequent chapter, *Americans Seek Justice*, we see how general societal division and disagreements over what should be taught on human origins have led to more than a century's worth of legal conflict. Here we see how individuals and organizations across America meet in court to go to battle over what should be taught to the next generation and what 'religious freedom' really means within the framework of a public-school classroom.

Once a general understanding of the complexity of the creationism | evolution conflict and its immense impact on education and society has been established, the analysis is taken a step farther to examine how such anti-evolutionary thought processes begin, how they are perpetuated and how they affect an individual's overall potential to become scientifically literate in *Morphology of Misconceptions*. Here a categorization of misconceptions and an overview of creationist thought is offered with a clear link to its effect on science education. Here we see how different degrees of literalism and identification with these literalist belief systems can have varying degrees of influence on a person's potential scientific literacy. As a person begins to identify more strongly with a particular misconception, their investment in upholding this belief becomes stronger. At the same time, belief systems that are based on very literal interpretations of the Bible are very rigid and restricting which acts as a major cause for a person to wholly reject the theory of evolution.

After establishing the cultural, educational and societal aspects of the creationism | evolution conflict, I put Theodosius Dobzhansky's 1973 claim "Nothing in Biology Makes Sense Except in the Light of Evolution" to the test in the chapter *Evolutionary Causes of Creationist Clinging*. Here we explore the potential evolutionary causes for creationist clinging and the rejection of evolution by examining the psychological, neurological and evolutionary causes driving the conflict even farther beyond its cultural context. The section title of this first part of the book, *Fight or Flight,* is actually intended to highlight the role of our own biology and evolutionary past in the creationism | evolution conflict. As we will discuss in this section, religion, religious beliefs and even the clinging to literalist interpretations of the Bible all make sense in the context of evolutionary psychology due to their role in the formation of 'meaning systems', their role in supporting social cohesion and the management of death anxiety. Because these religious beliefs play such a major role in people's personal lives and their understanding of the world, it becomes more understandable how evolution is perceived as threat and can trigger a fight-or-flight response, characterized by argumentation, rejection or cognitive shut-down.

This analysis and understanding of the influence of our own evolutionary history on our modern interpersonal relations allows us to then better recognize that while certain behavioral attributes, such as the ability to quickly identify potential threats and tendency to distrust 'others', certainly offered survival advantages to our primitive ancestors, these same tendencies are now a major driving force behind the distrust and, at times, outright hostility, between the major players in the evolution | creationism conflict.

It is irrefutable that the topic of evolution is an inherently emotional topic for many creationists and our entrenchment in an 'us vs. them' situation only exacerbates the emotional saliency of this topic. It is therefore imperative, not only that we recognize the negative influence of primitive behavioral tendencies but that we also learn to move beyond them if we are to make any educational or societal headway. To accomplish this, we need to first understand how our so-called 'downstairs brain' is affected by our evolutionary past and how these ancient neural structures influence our views, reactions and behavior—especially in highly emotional situations— so that we can then learn to engage our more highly evolved neural circuitry, i.e., our 'upstairs brain'. Understanding the role and structure of the 'downstairs brain' enables us to comprehend the (at times) hostile reactions displayed by those individuals who are emotionally invested in their beliefs about human origins or whose identity is closely connected to those beliefs (religious or materialistic). Once we have recognized how we are influenced by our biological past, we can move on literally and figuratively to *Tend & Befriend.*

The second half of the book focuses on finding solutions to the conflict over human origins with the ultimate goal of increasing scientific literacy potentials and simultaneously decreasing societal division. While the title of the first half, *Fight or Flight*, was used to highlight the negative influence of our primitive neurological circuitry on the creationism | evolution conflict, the title of the second half, *Tend & Befriend*, is used to highlight the potential of resolving this conflict using the more highly evolved portions of our brain.

While it is obvious that creationism, as an anti-scientific movement, must be vehemently opposed—especially when attempts are made to remove evolution from state standards or to replace it with 'alternative theories' such as intelligent design—I am also convinced that we must differentiate between creationism as a movement and the creationist as an individual. The same type of aggressive protection that is required at a legislative level to secure the quality of science standards is counterproductive at a classroom level as this type of fierce confrontation is likely to only enforce creationist students' fears of science and act as verification of their perceived conflict between science and faith. Creationist students, in particular, need to be recognized as a product of this movement and not the driver of it.

When looking for solutions, we must accept that there is no simple solution to such a complex problem. It is undeniable that anti-science misconceptions can negatively affect an individual's scientific literacy potential, and that a person's ability to understand and accept scientific principles is deeply affected by their upbringing, their peers and their worldviews. For that reason, anti-scientific or anti-evolution views cannot be simply overridden or undone by merely presenting clear data on evolution. It is therefore unrealistic to expect a creationist to change their views after one conversation or one lesson on evolution. These changes occur slowly (if at all) and may take years to develop. To support this change, we need concrete teaching methods and communication strategies that take the true complexity of this issue into consideration and thus address this issue at the most crucial level—namely at the level of cognition. While changing cognitive structures requires much more time than what is offered for a typical biology lesson or televised debate, it is possible to plant a seed of scientific interest in these moments that can then germinate over time and possibly take root much later in life.

The immense time required for conceptual change to occur in the case of creationism is based on the fact that creationist beliefs are deeply integrated into a person's self-theory and world-theory. Getting a student to understand and accept evolution, when they already maintain previous creationist beliefs inevitably requires major restructuring of the student's understanding of themselves and their role in the world. This level of conceptual change cannot be underestimated nor taken lightly as it requires a large degree of willingness and motivation for an individual to reconsider and rework previously held beliefs. While accepting evolution will require very literal creationists to rework much of their understanding of the natural world, we must guard against creating a dichotomous view where they feel as if they must choose between science and religion, or evolution and creationism. This is particularly true in educational settings, where students must be given the option to become scientifically literate while still maintaining their own worldviews, meaning systems and beliefs. The methods presented in the second half of the book will thus focus on increasing mental flexibility and well-being simultaneously in order to achieve the highest learning potential.

The second half of the book begins with the chapter *All in the Mind* where we take a general look at the idea of conceptual change and the role of emotions in learning to better appreciate the complexity of this educational challenge. We also take a special look at the architecture of the adolescent brain because evolution is most intensely taught during high school and it is therefore useful to develop an understanding of the neurological make-up of teenaged students. It is important to appreciate the intense role of emotions during this developmental stage and to recognize the unique learning potentials that exist at this age.

In *Conscious Evolution*, we look at how cognitive tools and practices, such as mindfulness, can be used to guide changes in our mental modes, essentially becoming the driver of our own conscious evolution. These practices help individuals become more aware of their own thought processes and emotional reactivity to those thoughts. By taking on the role of the observer, individuals, i.e., students, teachers, creationists, new atheists,[1] etc., attain more flexibility and control in their choice of reactions. As we will discuss in *Education (R)evolution*, these general tools can be applied to the creationism | evolution conflict to allow individuals to mindfully recognize the role of thoughts, judgments and emotions in their rejection of evolution, ultimately affording them more freedom in their choice of responses, i.e., to listen to the data presented to them instead of antagonizing the teacher. Likewise, new atheists may recognize their tendency to judge creationists as ignorant and then also achieve enough mental space from their judgmental thoughts to actively choose whether or not they insult creationists or if they engage with them in a productive discourse.

In a perfect world, an individual could also use these cognitive tools to develop enough insight into their own thinking patterns to be able to distinguish between fundamentalist rigidity and religious faith. In this way, even fundamentalist student could disengage from their need to uphold the literal interpretation of Genesis, for example, and thus become more responsive to lessons on evolution while still maintaining their faith. Unfortunately, it still remains difficult for many fundamentalists to incorporate science and religious beliefs and thus they will either reject science or suffer a full loss of their faith. In this case, these cognitive tools can also be used to deal with the painful emotions that arise during discussions of human origins, especially when these lessons do cause students to question the literal foundation of their belief system.

Finally, in *Understanding Common Descent* we look at ways to reduce the division around this topic by promoting understanding, tolerance, empathy and compassion. Many accusations have been levied against evolution, such as: it causes a materialistic look on the word; it leads to atheism and immorality; it promotes a harsh, animalistic approach to life based on competition and a struggle for survival, etc. Yet, if we look at the basic tenet of evolution, i.e., the concept of common descent, we actually find an idea that promotes one of the most inclusive and harmonious views on the world, e.g., we are all intrinsically connected. In this section we examine how we can reduce our tendency to 'other' and instead seek out our common humanity. In other words, we can see the creationist independent from their creationist views or the atheist independent from their atheist views. Ideally, through the development of more emphatic practices, it will be possible to decrease the general division in society, bridge societal gaps and to establish more positive learning environments. These conditions are a necessary prerequisite for effective dialogues that can then take place even between individuals who hold vastly opposing viewpoints.

In writing this book, I hope to find a means of increasing scientific literacy and of decreasing the general level of hostility that surrounds the discussion of human origins. Ultimately, I believe that both objectives can be accomplished using the same means—namely by understanding our own evolutionary history and potential for continued cognitive development so that we can learn to engage with one another using our highest cognitive functions. In other words, the key to solving this conflict lies in moving from primitive brain functions to more highly evolved mental modes and social discourse.

---

[1] The term 'new atheists' was coined in 2006 by *Wired Magazine* author Gary Wolf to describe active opponents of religion and spirituality such as Richard Dawkins, Sam Harris, and Daniel Dennett.

# Part I
# Fight or Flight

*Oxford Dictionary:*

*fight or flight: The instinctive physiological response to a threatening situation, which readies one either to resist forcibly or to run away.*

*Wikipedia:*

*The fight-or-flight response (also called hyperarousal, or the acute stress response) is a physiological reaction that occurs in response to a perceived harmful event, attack, or threat to survival. It was first described by Walter Bradford Cannon. His theory states that animals react to threats with a general discharge of the sympathetic nervous system, preparing the animal for fighting or fleeing.*

# Chapter 1

# Religion Meets Science

A recent question on 'Research Gate' caught my attention. One researcher asked, "Is it possible me to be a scientist and still maintain my religious views and belief in God?" The question received a myriad of answers, often contradictory. The answer to the question seemed simple enough to me. I grew up in a predominantly Catholic community on the border to Mexico, where many of my teachers and professors at university as well as my own mom were simultaneously scientists as well as devote Catholics. But this compatibility between religious beliefs and scientific careers is not so self-evident for many individuals who adhere to more fundamentalist belief systems. Here I will examine this general question to see what conflict, if any, exists between religion and science and whether this conflict is inherent or artificially constructed. Is it possible for religion to co-exist with science? Does an individual have to choose between their God and their interest in science? In this Section I will discuss the current research in these areas and offer insights by comparing various religions stance on evolution and take a specific look at the differences between traditional mainline Christianity and evangelical Christianity to determine how these differences affect adherents' ability to understand and accept scientific tenets.

## Christianity and the Fundamentals

The question of whether it possible for a person to believe in God and to accept the theory of evolution or are they mutually exclusive is not a novel question and has already been addressed many times by scientists and religious leaders alike. Science philosopher, Michael Ruse explicitly stated that this struggle is more legend than truth (Ruse 2001), while Stephen Jay Gould, American paleontologist, evolutionary biologist, and historian of science, has also vehemently proclaimed that there is an absolute lack of conflict due to the two very different realms of religions and science (1997). Even Pope Benedict XVI and his predecessor Pope John-Paul II have both praised the role of science in the development of humanity and acknowledged the strength of the theory of evolution, which has freed Catholics from seeing any conflict between their belief system and scientific progress (Numbers 1998).

Yet, there are those who see a clear and threatening conflict between their religious views and the theory of evolution and who continue to vehemently oppose the teaching of evolution. The driving force behind this opposition to evolution is the idea that evolution contradicts the biblical account of special creation in Genesis and that this contradiction can cause a loss of faith (Ham 2012; Morris 1989). It cannot be denied that evolution conflicts with the *literal* account of the Genesis story. But the question then becomes—was the Genesis story ever meant to be read literally? Was it ever meant to act as a how-to book on the creation of the world? Or was it perhaps meant for something else—such as a means of describing the nature of God and people's relationship to him?

The importance of discussing whether or not the Bible was written in order to be interpreted literally is important because this idea has been at the center of all creationist accusations against the teaching of evolution. According to Ken Ham,[2] Genesis forms the foundation of Christianity and if Genesis were to be lost, Christianity would tumble (2012). Yet, if we look back in the time to the point in history when Christianity was on its way to becoming a major world religion, we find St. Augustine[3] who argued *against* the literal interpretation of biblical texts. As St. Augustine explained, the Bible was written in a language that should be understood by relatively uneducated people since this was the characteristic of the mass population at the time that the Bible was revealed to human kind (Dixon 2008). This idea is known as the principle of accommodation or condescension and according to this theological principle, God still maintains His divine nature, yet it is acknowledged that His message has been related to humanity at the original audience's general language and education level so that humans can easily understand it (McGrath 1998). According to this theological principle Genesis does not need to be read as a literal account of the creation of the Earth for it to provide a foundation of the Judeo-Christian belief system that revolves around the concept of a single, almighty, omniscient God. When Genesis is read in this manner, it no longer conflicts with evolutionary theory, and it becomes much easier for religious individuals to maintain their faith while simultaneously embracing science.

Conrad Hyers takes this idea one step further and proposes that the original purpose of the Genesis account was allegorical. According to Hyers the Israelites used Genesis as a means of describing the true, all-powerful nature of the Hebrew God and to declare that their God was superior to the summation of gods worshipped by the Egyptians or Babylonians. According to this allegorical interpretation, each day of creation dismisses a group of gods or deities, for example, the statement "And God said 'Let there be light'... . And God called the light Day, and the darkness He called Night" (Genesis 1:3) was a means of saying that their God of Abraham was higher (more powerful/more real) than all of the pagan gods of light and darkness (Hyers 1984). In the same way, when "God made the firmament, and divided the

---

[2]   Ken Ham is founder of the Creation Museum and the Ark Encounter. He is also president of Answers in Genesis (AiG).

[3]   Augustine of Hippo (354–430 CE) was one of the most important Church Fathers in Western Christianity due to his work during the Patristic Era.

waters…" (Genesis 1:7) the Hebrew's God displaced the pagan gods of the sky and the seas. This pattern then continues throughout the rest of the week of creation, vanquishing the pagan gods while their God was established as superior to the sun, the moon and the stars. By placing their God over the gods of the Egyptians and Babylonians, the Hebrews were also able to call the divinity of the kings and pharaohs into question.

The beauty of the allegorical nature of the Bible is that it allows for the compatibility of religion, faith and science as described by Pope John Paul II. According to Pope John Paul II, who was declared a saint by the Vatican in 2014, the crux of the biblical account of creation does not lie in the details of the literal interpretation of the creation of the universe but instead in the understanding of the relationship between man, God and the universe as he said in his 1981 address to the Pontifical Academy of Science, "The Bible itself speaks to us of the origin of the universe and its makeup, not in order to provide us with a scientific treatise but in order to state the correct relationships of man with God and with the universe. Sacred Scripture wishes simply to declare that the world was created by God, and in order to reach this truth it expresses itself in the terms of the cosmology in use at the time of the writer." Pope Benedict XVI made a very similar statement, explaining that "The story of the dust of the earth and the breath of God, which we just heard, does not in fact explain how human persons come to be but rather what they are… . And vice versa, the theory of evolution seeks to understand and describe biological developments… . To that extent we are faced here with two complementary—rather than mutually exclusive—realities (Ratzinger 1995)." These types of statements made by the leaders of the Catholic church have prevented Catholics from having to choose between their faith and scientific fact, as H.L. Mencken stated "[The advantage of Catholics] lies in the simple fact that they do not have to decide either for Evolution or against it. Authority has spoken on the subject; hence it puts no burden upon conscience, and may be discussed realistically and without prejudice (1925)."

According to Ontario Consultants on Religious Tolerance, while most Catholics and liberal Protestants are easily able to reconcile their faith and science in their acceptance of either a theistic (guided by God) or naturalistic (unguided) view of biological evolution, the majority of evangelical Protestants continue to believe in the literal truth of the stories of creation found in the book of Genesis in the Hebrew Scriptures—interpreting the Hebrew word 'Yom' as meaning that creation took six actual 24-hour days. Current promoters of a literal interpretation of the Bible continue to purport that these texts *should* be interpreted as a verbatim account of God's exact actions in the creation of the universe (Ham 2012, 2013; Morris 1961, 1974). According to Hemminger, this insistence on a literal interpretation of biblical accounts is the root of the strife between religious and scientific communities (2009).

In order to ameliorate this problem, clergy members and rabbis across America have banded together to help spread their pro-science message in the form of an open letter (see theclergyletterproject.org). The open letter has already been signed by almost 15,000 American clergy members (as of December 2018) from different Christian denominations affirming the compatibility of Christian faith and the teaching of evolution (Figure 2). A similarly-worded letter has also been written and signed by Rabbis, Buddhist leaders and Unitarian Universalists.

---

### The Clergy Letter—from American Christian Clergy
#### An Open Letter Concerning Religion and Science

Within the community of Christian believers there are areas of dispute and disagreement, including the proper way to interpret Holy Scripture. While virtually all Christians take the Bible seriously and hold it to be authoritative in matters of faith and practice, the overwhelming majority do not read the Bible literally, as they would a science textbook. Many of the beloved stories found in the Bible—the Creation, Adam and Eve, Noah and the ark—convey timeless truths about God, human beings, and the proper relationship between Creator and creation expressed in the only form capable of transmitting these truths from generation to generation. Religious truth is of a different order from scientific truth. Its purpose is not to convey scientific information but to transform hearts.

We the undersigned, Christian clergy from many different traditions, believe that the timeless truths of the Bible and the discoveries of modern science may comfortably coexist. We believe that the theory of evolution is a foundational scientific truth, one that has stood up to rigorous scrutiny and upon which much of human knowledge and achievement rests. To reject this truth or to treat it as "one theory among others" is to deliberately embrace scientific ignorance and transmit such ignorance to our children. We believe that among God's good gifts are human minds capable of critical thought and that the failure to fully employ this gift is a rejection of the will of our Creator. To argue that God's loving plan of salvation for humanity precludes the full employment of the God-given faculty of reason is to attempt to limit God, an act of hubris. We urge school board members to preserve the integrity of the science curriculum by affirming the teaching of the theory of evolution as a core component of human knowledge. We ask that science remain science and that religion remain religion, two very different, but complementary, forms of truth.

---

**Figure 2:** The Christian clergy letter from the Clergy Letter Project aimed at countering the myth of conflict between religion and science.

In reading the letter from the American clergy, we can see two things: (1) teachers and scientists are not the only ones who are concerned about the creationism being taught as science, and (2) the impetus for battle between faith and science has not come from the church, theologians or the clergy, but was instead born outside of mainstream religion within a branch of Christianity known as conservative evangelicalism or fundamentalist Christianity.

## American Fundamentalism

According to George Marsden, "…an American fundamentalist is an evangelical who is militant in opposition to liberal theology in the churches or to changes in cultural values or mores, such as those associated with 'secular humanism'." In *Understanding Evangelicalism and Fundamentalism*, Marsden also points out that although the term 'fundamentalist' is often used today to describe any militantly traditionalist religion, such as Islamic fundamentalism, the term 'fundamentalist' was actually invented in the US in the 1920s to describe militant evangelicals who wanted to return Christianity to its fundamental teachings. This is reflected in the current definition of fundamentalism from the Oxford dictionary:

> *Fundamentalism: (noun) 1. A form of a religion, especially Islam or Protestant Christianity, that upholds belief in the strict, literal interpretation of scripture: Modern Christian fundamentalism arose from American millenarian sects of the 19th century, and has become associated with reaction against social and political liberalism and rejection of the theory of evolution.*

In looking at the origins of Christian fundamentalism we see that it began as a systematic theology by the 1920s within the Protestant churches. American fundamentalists started the publication of a series of appropriately entitled booklets called *The Fundamentals*. If one looks at an essay from these booklets, such as *A Testimony to the Truth*, we see that the point of the essay is aimed at defending Protestant orthodoxy while attacking such topics as (1) higher criticism,[4] (2) liberal theology, (3) socialism, (4) modern philosophy, (5) atheism, (6) Catholicism and (7) evolutionism.

Social changes of the early twentieth century also fed the flames of fundamentalism as "Fundamentalists felt displaced by the waves of non-Protestant immigrants from southern and eastern Europe flooding America's cities. They believed they had been betrayed by American statesmen who led the nation into an unresolved war with Germany, the cradle of destructive biblical criticism. They deplored the teaching of evolution in public schools, which they paid for with their taxes, and resented the elitism of professional educators who seemed often to scorn the values of traditional Christian families (Wacker 2000)."

The ideas of an inerrant and infallible Bible that should be read as a literal account also arose during this time as a part of the fundamentalist reaction to liberalism and rationalist thought. Benjamin B. Warfield, one of the authors of *The Fundamentals*, is credited with the advancement of the concept of biblical inerrancy (Orr et al. 1910) as stated in blurb on the 2008 reprint of Warfield's 1927 essays: "B.B. Warfield's volume on divine revelation and biblical inspiration defined the parameters of the twentieth century understanding of biblical infallibility, inerrancy, and the trustworthiness and authority of Scripture."

This idea of an infallible Bible caught on quickly and became highly popular as it was perpetuated by influential seventh-day Adventists George McCready Price, John C. Whitcomb, Jr. and Henry Morris. Like other fundamentalists of the early 1900s, Price also believed that evolution would lead to the demise of Christianity, ethics and political freedom. He was convinced of the Bible's infallibility and aimed to find scientific data to support biblical accounts (Numbers 1992). Price produced a manuscript for a book in 1902, preceding the publication of *The Fundamentals*, in which he proposed that there was geological evidence of the story of Genesis. He explained that the sequence of fossils resulted from the different responses of animals to the encroaching flood and how Niagara River Gorge, the Grand Canyon, the Alps and the Himalayas had also been formed during this great flood (Numbers 1992). By 1923 he had created his own college textbook entitled *The New Geology* in which he put forth these ideas and ultimately sold over 15,000 copies of his book (Numbers 2006). Initially, Price's interpretation of geological history, which was based on the literal interpretation of the Bible, remained limited to the peripheral groups of seventh-day Adventists, but his ideas become popular with mainstream America soon after Price began to collaborate with John C. Whitcomb, Jr. and Henry Morris (Numbers 2014).

---

[4]   Higher criticism is a branch of literary criticism which analyzes ancient texts in order to understand 'the world behind the text' and it is based on the idea of rationalism—a belief or theory that opinions and actions should be based on reason and knowledge rather than on religious belief or emotional response.

Together with a group of Adventists, Price founded the Deluge Geology Society (DGS) in 1938. Members of the DGS were required to believe in the six literal days of Creation and to devote themselves to the study of the Deluge as the major cause of the geological changes since creation (Numbers 1992). This society provided the link between Price and Henry Morris, who became a member of Deluge Geology Society while he was a graduate student. Morris had chosen to study hydraulic engineering with a minor in geology so that he would have a good understanding of how the flood waters had shaped the face of the Earth and took the idea of diluvial geology and ran with it (Flank 2007).

Like Price, Henry Morris clearly distained evolution, as he stated later in his life, "Evolution is the root of atheism, of communism, Nazism, behaviorism, racism, economic imperialism, militarism, libertinism, anarchism and all manner of anti-Christian systems of belief and practice (1972)." Yet unlike Price, Morris was able to bring seventh-day Adventists beliefs to mainstream America. While Morris was not the first to claim scientific rationale for a young Earth, he was the first to be able to draw a large following. Numerous historians argue that Morris was one of the most influential creationists in the second half of the twentieth century as he is seen as the figure responsible for moving the majority of creationists away from accepting the old age of the Earth and towards a more literal interpretation of the Bible (Numbers 2006).

Morris published hundreds of books and pamphlets. He co-authored his most well-known publication *The Genesis Flood*, which was published for the first time in 1961, with John C. Whitcomb, Jr. Although their original work had been greatly motivated by the work done by Price, Morris and Whitcomb deleted almost all mention of Price and the Adventists from the book in the hope of a new start and more receptivity to their ideas among mainstream America (Numbers 1992).

According to Matzke, *The Genesis Flood* is the most important creationist book of the 20th century since the book and Morris "transformed [young Earth creationism] from a somewhat obscure doctrine of extreme fundamentalists and spread it far and wide across evangelical churches (2010)." This book did indeed start a revolution as Michael D. Gordin, science historian, describes it, *Genesis Flood* is "one of postwar America's most culturally significant works about the natural world. It was read by hundreds of thousands, spawned its own research institutes, and remains absolutely rejected by every mainstream biologist and geologist (2012)."

The research institutes that Gordin was referring to are the Creation Research Society (CRS) founded by Morris in 1963, the Creation Science Research Center (CSRC) founded in 1970 and the Institute for Creation Research (ICR) founded in 1972. Like the DGS, these societies also required members to believe in an infallible Bible and the Creation Research Society even created a formal statement of belief that members were required to sign (Figure 3).

The centrality of this idea of an inerrant Bible flourished in the United States, especially during the 1970s and 1980s. The International Council on Biblical Inerrancy (ICBI), for example, was established in the United States in 1977 "to clarify and defend the doctrine of biblical inerrancy". In 1978 they published the Chicago Statement on Bible inerrancy in which they stated:

---

**Creation Research Society Statement of Belief (Numbers 1992)**

1. The Bible is the written Word of God, and because it is inspired throughout, all its assertions are historically and scientifically true in all the original autographs. To the student of nature this means that the account of origins in Genesis is a factual presentation of simple historical truths.
2. All basic types of living things, including man, were made by direct creative acts of God during the Creation Week described in Genesis. Whatever biological changes have occurred since Creation Week have accomplished only changes within the original created kinds.
3. The great Flood described in Genesis, commonly referred to as the Noachian Flood, was an historic event worldwide in its extent and effect.
4. We are an organization of Christian men of science who accept Jesus Christ as our Lord and Saviour. The account of the special creation of Adam and Eve as one man and woman and their subsequent fall into sin is the basis for our belief in the necessity of a Saviour for all mankind. Therefore, salvation can come only through accepting Jesus Christ as our Saviour.

---

**Figure 3:** Creation Research Society's membership statement of belief emphasizing the literal interpretation of the Bible.

*"The authority of Scripture is a key issue for the Christian Church in this and every age. Those who profess faith in Jesus Christ as Lord and Savior are called to show the reality of their discipleship by humbly and faithfully obeying God's written Word. To stray from Scripture in faith or conduct is disloyalty to our Master. Recognition of the total truth and trustworthiness of Holy Scripture is essential to a full grasp and adequate confession of its authority. The following Statement affirms this inerrancy of Scripture afresh, making clear our understanding of it and warning against its denial."*

This focus on the infallibility of the Bible and the necessity to accept the Genesis account as the literal description of the beginning of the universe has forced many such adherents to choose between their faith or science. As stated above, "To stray from Scripture in faith or conduct is disloyalty to our Master". But must all religious individuals choose between their God and science? According to the results of a PEW poll—that answer is no (Figure 4).

These results were obtained as part of a study conducted by the Pew Research Center in 2008 regarding religious affiliation and acceptance of evolution. In their study they posed the question of whether or not participants believed that evolution is the best explanation for the origin of human life on Earth. Their results show a clear delineation of religious membership and adherents understanding and acceptance of evolution. While Jehovah's Witnesses have the lowest acceptance of evolution with roughly only 8% believing that evolution was the best explanation of human origins, they are closely followed by with Mormons (22%) and Evangelical Protestants (24%) who also well below the national average (48%). Leading the nation regarding the acceptance of evolution were the Buddhist (81%), Hindu (80%) and Jewish (77%) adherents, who were not only above the national average but also all showed a higher degree of acceptance of evolution than even those individuals who were religiously unaffiliated. This study highlights the fact that there is not a necessary conflict between science and religion and that a wide spectrum of acceptance of evolution exists within Christian groups with Catholics showing the highest amount of acceptance with 58%, followed by Orthodox with 54% and Mainline Protestants with 51%. Here it should be

% who agree that evolution is the best explanation for the origins of human life on earth

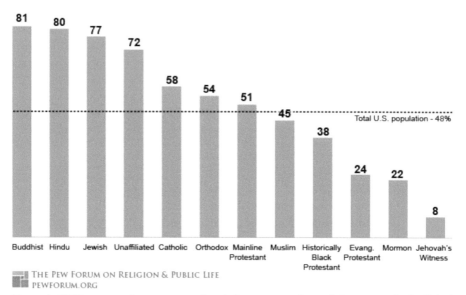

Figure 4: PEW report on the acceptance of evolution among various religious groups in the United States.

noted that in terms of science, evangelical Protestant views are much closer to those of Mormons and Jehovah's Witnesses than to mainline Protestants.

So, to return to the original questions: Is it possible to have a relationship to God and still accept scientific facts? Or to even be a scientist? My humble answer to this question on 'Research Gate' was—yes, of course—so long as your faith is not fundamental in nature, i.e., not dependent on the literal reading of a religious text. Yet, unfortunately the current trend in America is moving farther towards this fundamentalist area as evangelical Protestantism now represents the largest religious fraction in the United States (PEW Research Center 2008). But what is it about America that causes there to be such a high prevalence of these particularly conservative and literal beliefs? In the next section we will examine certain aspects of American history and culture that may explain why creationism took root and has continued to flourish in the United States.

## Religious America

*Creationism is a homegrown phenomenon of American sociocultural history*
Stephan Jay Gould (1997)

While creationism and intelligent design ideas are present in many countries, they originated in the United States and are particularly popular among Americans— in fact, roughly half of the general population harbors significant sympathy for creationism and 25% is actually fundamentally opposed to evolution (Matzke 2010). Why America? Why is there such a prevalence of evolution-doubters in the United

States? How and why did fundamentalism and evangelicalism develop in the United States?

According to historian Stephen Prothero, there has been this idea that America is a holy land since the very beginning of American life, "We've had this notion that this is a special place, and what makes it special is that we have some kind of special relationship with God (2012b)." When we look back at the early settlers of the US, we find the Winthrop Fleet, which brought eleven ships to Massachusetts in 1630. Over 1000 Puritans listened to Winthrop's sermon upon his ship entitled 'A Model of Christian Charity'. This sermon contained the idea that they were en route to create a new society, 'a city on a hill'. This concept of a city on a hill comes from Jesus' Sermon on the Mount in Matthew 5:14, "You are the light of the world. A city that is set on a hill cannot be hidden." This speech reflected the Puritan's beliefs about the importance of their migration across the Atlantic. "In the eyes of the Puritan leaders the settlement of New England appeared to be the most significant act of human history since Christ bade farewell to His disciples...An entire community living as God had directed men to live—this was the vision that impelled thousands of people to cross the Atlantic (Miller 1966)." The Puritans brought with them a story which told them that they had left Europe in order to remake Christianity and this gave them a sense that they were the chosen people with a special purpose (Prothero 2012a).

In addition to the Puritans, the Quakers also had a large effect on the early religious landscape of the colonies. While the Puritans wanted to bring Christianity back to its purist form based on the Bible, the Quakers attempted to remove all intermediaries between God and his people. The Quakers were important members of the establishment of eight English colonies in North America (Miller 1966) and William Penn was one of their strongest leaders. William Penn established West Jersey, Delaware and Pennsylvania with the plight of the Quakers in mind and hoped that he would be able to "reproduce true primitive Christianity (Miller 1966)." For the Quakers these establishments were like a 'holy experiment' much like the concept of a 'city on the hill' in New England (Barbour and Frost 1998).

Fast forward one hundred years and we find America entrenched in the Great Awakening which took place in the 1730s and 1740s. The Great Awakening was an evangelical movement characterized by powerful sermons that gave the listeners a sense of a deep and personal relationship to their Lord Jesus Christ and emphasized the need for personal salvation while also fostering a deep sense of spiritual conviction and a commitment to personal morality (Kidd 2007). One of the most influential leaders during this time was Reverend George Whitefield who began his preaching tour of the American colonies in 1739 after experiencing a 'spiritual rebirth' as he liked to call it, which was a transformation of the soul by the Holy Spirit of God. According to *Christianity Today*, Whitefield gave 18,000 sermons and spoke to over 10 million people during his lifetime was "probably the most famous religious figure of the eighteenth century". Whitefield's concept of a religious 'rebirth' changed the face of religion in the US and quickly became very popular among the American people (Belton 2012a).

This idea of a personal, direct and zealous relationship to God only continued to grow. As America began to expand westward, more and more Americans moved to the western frontier. While the western frontier offered many opportunities for expansion, there were very few established churches in these new territories. This lack of religious authority meant that the Bible was being read at face value by fairly uneducated individuals (Matzke 2010) and revivals began to dominate as the religious gatherings of the west (Belton 2012b).

These revivals took place in clearings in the woods and drew large crowds. The revival in Cane Ridge, for example, drew masses of 20,000 people in 1801. Charismatic ministers warned the crowds of the new nation's spiritual crisis, and there was a "sense of urgency in trying to bring people to Christ because they're not just saving them, they're saving the nation, as well (Lyerly 2012)." The popularity of revivals increased exponentially all across the union and by 1811 more than 1,000,000 Americans were visiting at least one religious revival per year (Belton 2012b). The Methodists organized circuits with traveling ministers, known as circuit riders, which was a perfect solution for the frontier lifestyle and these preachers had access to many more people than a traditional preacher who stayed at one local church (Lyerly 2012). It is not surprising that the Methodists also received the most converts at this time and these rapid conversions also offered a great means of quick fundamental change. By the mid-1800s, the religious landscape of the United States had been completely transformed. While there had been 15,000 Methodist at the time of the Revolution, there were now more than 1 million. More Americans were attending church than ever before and two-thirds of them were evangelical Protestants.

It was also during this formative period in American evangelicalism that Darwin published his *Origin of a Species* (1859). Just as millions of Americans were attending revivals where they were learning about how Jesus will return to Earth to judge the wicked, a scientific theory emerged that challenged the core concept of the perspicacity of Scripture. Just two years later the idea of a 6000-year-old Earth was crushed by Lord Kelvin who estimated the Earth to be between 20 and 40 million years old.

As America moved into the 20th century, evangelicals felt as if they had lost hold of their holy nation and Christian fundamentalism arose as a reaction to a perceived threat to traditional Protestant beliefs and a concern about the role of Christianity in American culture (Longfield 2000; Numbers 1998; Scott 2009). One of the greatest threats to traditional religious beliefs appeared to be the general movement towards modernism as illustrated by E.J. Pace in his illustration entitled 'The Descent of the Modernist' which shows how the modernists were taking steps to atheism by questioning the infallibility of the Bible and man's role at the crown of God's creation (Figure 5).

This fear of modernism and specific trepidation regarding science and technology were exacerbated by the horrendous losses resulting from World War I as historian Randall M. Miller describes, "From the traditionalist point of view, this war was a demonstration of all that had gone wrong, and a warning because God, they believed, gives warnings. He visits his wrath upon the unrepentant people. The world seemed to be coming apart. How can we pull these things all back together (2012)?" Evangelicals were quick to see the connection between the evils of war and the teaching of evolution, as William Jennings Bryan stated, "The same science that manufactured poisonous gases to suffocate soldiers is preaching that man has a

**Figure 5:** The Descent of the Modernist by E.J. Pace (1922) illustrates the perceived development of atheism through modernism.

brute ancestry and eliminating the miraculous and the supernatural from the Bible (Numbers 1998)." Soon evangelicals began to feel as if they were losing their special charge over this chosen nation and that it was time to change that (Miller 2012).

More and more evangelicals began to refer to themselves as fundamentalists to highlight the idea of returning to the core of Christian teachings as there was a general sense that giving up the truth of the Bible was akin to giving up all of Christianity (Kazin 2012). As William Jennings Bryan said, "Man must be brought back to God, to a belief in the Bible as the Word of God, and to a love of Christ as the Son of God (Colt et al. 2012)." Historian, Hasia R. Diner, poignantly describes the crux of the problem, "A core religious belief was that human beings were the crown of creation. And in very American terms, the American was also the crown of creation. But now, reading these accounts of Darwin, one couldn't say that any longer. Darwinism undermined the notion of what it means to be an American (2012)."

So, Bryan and the fundamentalists began to campaign to have the teaching of evolution removed from American schools. Bryan proclaimed that Darwinism was dangerous for both religious and social reasons due to its reliance on naturalistic explanations for human origins. As he stated in 1904, "I object to the Darwinian theory, because I fear we shall lose the consciousness of God's presence in our daily life, if we must accept the theory that through all the ages no spiritual force has touched the life of man and shaped the destiny of nations (Larson 1997)." Bryan also specifically objected to the mechanism of natural selection, which he called the law of hate, "The Darwinian theory represents man as reaching his present perfection by the operation of the law of hate—the merciless law by which the strong crowd out and kill off the weak (Larson 1997)." This movement against evolution was very attractive among the evangelicals who saw evolution as a moral degradation and a threat to the traditional interpretation of the Bible (Blancke et al. 2014). This fear of evolution and the effect of science education on young impressionable minds was again captured in an illustration be E.J. Pace (Figure 6).

The fundamentalists' efforts were successful as twenty-three states debated anti-evolution legislation and three states passed legislation prohibiting the teaching of evolution (Tennessee, Mississippi and Arkansas). The prohibition in Tennessee led to the Scopes Trial in 1925. Although John Scopes was found guilty and the *Butler Act* was upheld in Tennessee, Bryan was made to look like a fool after being questioned by Darrow (Larson 1997). Bryan died shortly after the case and the fundamentalists retreated into the shadows. The massive media coverage of the case polarized the American people and there was a clear division between the liberals and the conservatives, between the modernists and the fundamentalists (Colt et al. 2012). According to historian Edward J. Larson, "The Scopes trial was such a visible repudiation of the fundamentalists by the mainstream media and mainstream culture that there was a sense that, 'Our ideas are no longer welcome. Rather than participating in the larger society, we should build our own subculture' (2012)." In a sense, the media coverage of the Scopes trial served as a vilification of the fundamentalists and thus during the 1930s and 1940s, the fundamentalists began to create a new image as part of their retreat—moving away from the term fundamental and focusing more on a cultural re-engagement of evangelicalism (Gribben 2011).

**Figure 6:** Illustration by E.J. Pace depicting the perceived loss of faith in God through the inclusion of evolution in public school education (From E.J. Pace's book 1922 book '101 Christian Cartoons').

## *Rise of The Religious Right*

Following the economic and political upheavals in the 1920s, 30s and 40s, a new America began to emerge where faith became equated with patriotism. This melding of Christianity and political activity was driven largely by the 'crusades' led by evangelist Billy Graham (Taddonio 2018). Graham's mix of media and live events made him one of the most influential religious figures in American history. The masses of people that were attracted to these 'crusades' made it possible for Graham to hold the record as the person who has preached the Gospel to more people in person than anyone in the history of Christianity (Horstman 2002). His influence over radio and television was even greater; in fact one particular TV 'crusade' reached an estimated 2.5 billion viewers worldwide (Stammer 1996). Graham was a sensational figure who made it onto Gallup's list of most admired men and women more than any man or woman in history (total of 61 times) (Newport 2018).

Not only did Graham have a major influence on the religious texture of the United States but also played an integral role in the blending of politics and evangelical beliefs. He has provided spiritual counsel to for *all* sitting presidents from Harry S. Truman to Donald Trump, and George W. Bush even credited Graham with his choice to "recommit [his] heart to Jesus (BGEA 2018)." Graham can also be credited with bringing the revival culture in America back to life and for starting a trend of political leaders placating to the evangelical masses (King 1997).

The 1950s marked a period of time in American history where the 'wall' between State and Church was apparently absent and instead, the word 'God' became more prevalent in government documents. The words 'under God' were added to the Pledge of Allegiance in 1954. In 1955, the New York Board of Regents decided to create a non-denominational prayer that was recommended to be recited by children in public schools in order to protect them against communist atheism. This prayer (later known as the Regent's prayer) read "Almighty God, we acknowledge our dependence upon thee and we beg thy blessings upon us, our parents, our teachers, and our country. Amen." One year later the phrase, "In God We Trust" became the official motto of the United States Congress and these same words were then added to American paper currency in 1957.

The number of the American evangelicals was growing so rapidly that it only made sense for the political leaders to appease them in order to remain popular among the majority of voters. By the mid-1960s Graham had become what Grant Wacker refers to as the 'Great Legitimator' because as Grant explains, "his presence conferred status on presidents, acceptability on wars, shame on racial prejudice, desirability on decency, dishonor on indecency, and prestige on civic events (2014, pp. 24–25)."

A new wave of evangelical activism arose in response to a series of Supreme Court Rulings, namely the outlawing of school prayer (*Engel v. Vitale* 1962; *Abington School District v. Schempp* 1963), the outlawing of statutes prohibiting the teaching of evolution (*Epperson v. Arkansas* 1967) and the legalization of abortion (*Roe v. Wade* 1973). In 1979, Jerry Falwell launched the Moral Majority, a political organization designed to bring evangelicals back into national politics. Within one year, the Moral Majority was set up in 47 states trying to mobilize 10 million evangelical voters.

This massive organization of like-minded individuals, who were interested in enacting and changing specific policies offered politicians a group that they can cater to in order to increase their chances of election. Presidential candidate, Ronald Reagan, recognized the potential power of appealing to this organized and mobilized group of voters, making a number of proclamations during the 1980 presidential campaigns that directly addressed the upsetting court rulings:

> *"Our positive stance on family and children is consistent with our heartfelt convictions on the issue of abortion. Here again, we are not just against an evil. We are not just anti-abortion. We are pro-life. In the meantime, we in government will see to it that not one tax dollar goes to encouraging any woman to snuff out the life of her unborn child."*

> *"No one will ever convince me that a moment of voluntary payer will harm a child or threaten a school or a state. But I think it can strengthen our faith in a creator who alone has the power to bless America."*

Ronald Reagan was successful in his campaigning and in his appeal to the evangelical Christians. The Moral Majority is accredited with securing Reagan's spot in the Oval Office with two-thirds of the white, evangelical population's vote (King 1997). According to Rev. Randall Balmer, Reagan's win came as a shock to many Americans and many became quite anxious about the power that the religious right had obtained in national elections (2012). In 1983, President Ronald Reagan reached back through hundreds of years of American history to dust off Winthrop's metaphor saying, "In the book of John is the promise we all go by, tells us that for God so loved the world that he gave his only begotten son, that whosoever believeth in him should not perish but have everlasting life. With his message and with your conviction and commitment, we can still move mountains. We can work to reach our dreams and make to America a shining city on a hill." According to Prothero and Rev. Balmer, this reference to the 'city on a hill' was a masterpiece of political symbolism that resurrected the idea of a Christian nation and a 'holy experiment'.

While these developments were clearly welcomed by the evangelical population, it was also very alarming to many American politicians. During that same year, Senator Ted Kennedy travelled to Liberty Baptist College to address the dangers of evangelicals attempting to influence government activity, "The separation of church and state can sometimes be frustrating for women and men of religious faith. They may be tempted to misuse government in order to impose a value which they cannot persuade others to accept. But once we succumb to that temptation, we step onto a slippery slope where everyone's freedom is at risk."

As opposition to the Moral Majority mounted, Jerry Falwell decided to dismantle the organization in 1989 (Utter and True 2004). In its place, a new form of political action was created by the religious right, led by former presidential candidate Pat Robertson and his Christian Coalition. The Christian Coalition is said to be the most powerful interest group within the Republican party and has received extensive credit for successes in Republican campaigning since its founding (King 1997). The Christian Coalition took a different approach than the Moral Majority by focusing on organizing evangelicals at the grass roots level, with a specific focus on school board elections (Barker and Belton 2012). Ralph Reed was hired to lead the efforts and he stated, "Religiously devout Christians are somewhere between 25 and 30 percent of the electorate. And I thought if we could figure out a way, by organizing them and mobilizing them and training them and deploying them and activating them, so that their influence and effectiveness was even proportional to their numbers, we would transform American politics (Barker and Belton 2012)." Reed and the Christian Coalition were successful in their efforts and by the 1990s, thousands of school boards across the country had conservative fundamentalist religious right majorities (Barker and Belton 2012).

The evangelical Christian political involvement continued into the 21st century and the movement gained new hope and momentum with the presidential elections in 2000. Bush expressed his Christian beliefs clearly and openly during the presidential campaigns, causing Rev. Pat Robertson to see the election of George W. Bush as the ultimate green light as an evangelical now held the most powerful position in the world. The recognition of power of organized voting blocks and grassroots actions among evangelicals continues to today. In fact, since the turn of the twenty-first century, the evangelical lobby in America has become the most powerful grassroots coalition in the country (Gribben 2011).

## Summary

Religion has played a major role in American culture since the first colonists arrived over 500 years ago. The earliest settlers brought with them the idea of creating a Christian nation and this idea has been passed down through hundreds of years of American history. Developments in the 1800s stirred up a new emotional element of American religiosity that was marked by a close and personal relationship with God.

As the history of the nation continued, this special relationship to God and the special role of religion in American's lives became coupled with political activism. In the 20th century, as the number of evangelicals began to rise exponentially, there was a major focus on organizing their effort to create political change. The success of these evangelical efforts offered politicians a new target group to court in order to improve their chances of political success, thereby giving the religious right unprecedented power in the United States that is still palpable today.

# Chapter 2
# Americans Seek Justice

While the previous chapter attempted to explain how and why creationism became so prevalent in the United States, it would be incorrect to assert that all Americans are religious zealots. In fact, there are arguably as many Americans who are avid secularist as there are religious fanatics. This can also be clearly seen in the statistics. While the emphasis of the statistics is on the fact that "only about half of Americans accept evolution as the most likely cause of human origins" the obvious flip-side of this statistical analysis is that the other half of the American population does accept the theory of evolution.

Nowhere is the existence of both the pro-secular and anti-secular fractions in America more clearly visible than in the legal cases involving evolution. As I am sure that every child has heard from their parents after tattling on a sibling—it takes two to fight—the same holds true in the legal system, i.e., for any type of court case to occur, there must be two opposing parties. The fact that there have been so many court cases surrounding evolution is not only a sign of the high level of activity among the anti-evolution fraction, but it also highlights the vigilance and dedication that has been exerted from the pro-science and pro-secular organizations to prevent these anti-scientific trends from taking root in the public-school system.

While the legal cases here describe many parents, who have acted as individuals in their attempt to preserve secular education, there are also a number of organizations who have been involved in many of these cases, acting as vigilant, science-advocating, secular watchdogs. Two of the most active and vocal of these organizations are the American Civil Liberties Union (ACLU) and National Center for Science Education (NCSE). While these two organizations have vastly different histories, they have one major common goal and that is the preservation of quality secular (science) education. It is therefore not surprising these organizations have played either a direct or indirect role in several of the cases that will be discussed in this chapter. Other organizations involved in the protection of secular science education include the National Association of Biology Teachers and the Americans United for Separation of Church.

# Secular America

Before beginning it should be clearly stated that the roots of this secular American thinking style embodied by these organizations is also deeply engrained in our country's history. Following the end of the Revolutionary War, the Founding Fathers were given a unique opportunity to create the framework of the new nation. The Founding Fathers created the Constitution of the United States to establish the manner in which the new nation should be governed. Here it appears that the Founding Fathers were more influenced by the ideas of enlightenment, rather than religious zeal. In contrast to the Declaration of Independence, all references to a Creator are absent in the preamble of the Constitution:

> *We the People of the United States, in Order to form a more perfect Union, establish Justice, insure domestic Tranquility, provide for the common defense, promote the general Welfare, and secure the Blessings of Liberty to ourselves and our Posterity, do ordain and establish this Constitution for the United States of America.*

Yet while the Constitution created a strong central government, the original document did not include any protection of individual rights. Thomas Jefferson believed that individual rights should be protected and incorporated into a part of the Constitution. The Bill of Rights was thus proposed in 1789. The First Amendment protects American citizens' freedom of religion, freedom of speech, freedom of press and freedom of assembly. This is the amendment most quoted and pivotal in all of the creationism/evolution court cases. It states:

> *"Congress shall make no law respecting an establishment of religion, or prohibiting the free exercise thereof; or abridging the freedom of speech, or of the press; or the right of the people peaceably to assemble, and to petition the Government for a redress of grievances."*

This simple statement was one of the earliest official proclamations of full religious freedom in the world. Yet, although it guaranteed that Congress would not establish any laws in favor of or in repression of a certain religion, it was not until Jefferson's presidency that he described the First Amendment as being a "wall of separation" between church and state. This description originates from a letter he wrote in 1802 regarding the First Amendment and it is the wording from this letter that is often quoted when the vagueness of the First Amendment causes issue in legal cases. In his letter, Jefferson stated:

> *"Believing with you that religion is a matter which lies solely between Man & his God, that he owes account to none other for his faith or his worship, that the legitimate powers of government reach actions only, & not opinions, I contemplate with sovereign reverence that act of the whole American people which declared that their legislature should 'make no law respecting an establishment of religion, or prohibiting the free exercise thereof,' thus building a wall of separation between Church and State."*
>
> Thomas Jefferson to Danbury Baptist Association 1802

The understanding of this separation of church and state and the extent of this separation is crucial for understanding the crux of evolution/creationist legal conflict. The idea of a wall of separation together with the First Amendment form the foundation for the rulings made in all of the legal disputes involving the teaching of creationism or creationist agendas in public schools.

The first part of the First Amendment about religious freedom is often broken down into two different clauses known as the 'Establishment Clause' ("Congress shall make no law respecting an establishment of religion…") and the 'Free Exercise Clause' ("Congress shall make no law…prohibiting the free exercise thereof…") in order to clearly distinguish the two aspects of religious freedom. While secularist rely most heavily on the Establishment Clause to keep creationist agendas out of the classroom, creationists continue to appeal to the Free Exercise Clause as an excuse to bring it back in.

Originally, the individual rights declared in the Bill of Rights only applied to the rights of the people in relation to the federal government, e.g., that Congress cannot establish religion. Since the public schools in the US are state institutions, they would have been exempt from having to uphold these rights under the original framework of the Bill of Rights, but the 14th Amendment, which was adopted in 1868, granted these rights to the people at a state level. This means that the state and all state institutions must also uphold the citizens' right to freedom from the establishment of religion. The fact that the schools are also bound by the Establishment Clause is what makes it possible for parents to sue the school district or the governor of a state for establishing creationist policies or conditions where creationism could be taught in public school classrooms. Without the 1st and 14th amendment, schools could legitimately teach creationism in classrooms instead of evolution. Yet, due to the Founding Fathers' emphasis on rationality and the separation of church and state, schools are now bound by law to provide secular education and the U.S. Supreme Court is especially vigilant in policing violations of the Establishment Clause within public schools for three main reasons: (1) students are required by law to attend school, (2) young children and adolescents are very impressionable by nature, and (3) teachers play an authoritative and influential role in the classroom (Wexler 2010).

## Conflicts in the Classroom

Despite the clear intention and meaning of the Establishment Clause, it has not prevented individuals from trying to enact laws and educational policies that clearly attempt to disassemble the wall between church and state. As discussed, there is a large and active population of religious right in the United States who are intent on having their religious beliefs permeate the texture of society. They are intent on affecting change to the societal structure and education system in order to have their religious views reflected and expressed in these arenas. Particular organizations, such as the Christian Coalition, focused specifically on affecting change within the public-school system. This type of movement and organizational efforts have led to the proposal and in statement of anti-scientific laws in multiple states and at multiple levels throughout the 20th century and continue to be proposed today. In Table 1,

**Table 1:** Overview of creationist/anti-evolution legislation from 1925, 1973, 1981 and 2001.

| Year | Law |
|---|---|
| 1925 | **Butler Act*** (1925): AN ACT prohibiting the teaching of the Evolution Theory in all the Universities, Normals and all other public schools of Tennessee, which are supported in whole or in part by the public school funds of the State, and to provide penalties for the violations thereof. <br> Section 1. Be it enacted by the General Assembly of the State of Tennessee, That it shall be unlawful for any teacher in any of the Universities, Normals and all other public schools of the State which are supported in whole or in part by the public school funds of the State, to teach any theory that denies the story of the Divine Creation of man as taught in the Bible, and to teach instead that man has descended from a lower order of animals. <br> Section 2. Be it further enacted, That any teacher found guilty of the violation of this Act, Shall be guilty of a misdemeanor and upon conviction, shall be fined not less than One Hundred ($ 100.00) Dollars nor more than Five Hundred ($ 500.00) Dollars for each offense. <br> *Tenn. HB. 185, 1925 |
| 1973 | **Genesis Act**** (1973): Any biology textbook used for teaching in the public schools, which expresses an opinion of, or relates a theory about origins or creation of man and his world shall be prohibited from being used as a textbook in such system unless it specifically states that it is a theory as to the origin and creation of man and his world and is not represented to be scientific fact. <br> Any textbook so used in the public education system which expresses an opinion or relates to a theory or theories shall give in the same textbook and under the same subject commensurate attention to, and an equal amount of emphasis on, the origins and creation of man and his world as the same is recorded in other theories, including, but not limited to, the Genesis account in the Bible. <br> ...Each school board may use textbooks or supplementary material as approved by the State Board of Education to carry out the provisions of this section. The teaching of occult or satanical beliefs of human origin is expressly excluded from this Act. <br> **1973 Tenn. Pub. Acts, Chap. 377 |
| 1981 | **Balanced Treatment for Creation-Science and Evolution-Science Act*** (1981): "Public schools within this State shall give balanced treatment to creation-science and to evolution-science." Section 4 Definitions, as used in this Act: <br> (a) "Creation-Science" is defined as scientific evidences for creation and related inferences that indicate: (1) Sudden creation of the universe, energy, and life from nothing; (2) The insufficiency of mutation and natural selection in bringing about development of all living kinds from a single organism; (3) Changes only within fixed limits of originally created kinds of plants and animals; (4) Separate ancestry for man and apes; (5) Explanation of the earth's geology by catastrophism including the occurrence of a worldwide flood; and (6) A relatively recent inception of the earth and living kinds. <br> (b) "Evolution-Science" is defined as being scientific evidences and related inferences that indicate: (1) Emergence by naturalistic processes of the universe from disordered matter and emergence of life from nonlife; (2) The sufficiency of mutation and natural selection in bringing about the development of present living kinds from simple earlier kinds; (3) Emergence by mutation and natural selection of present living kinds from simple earlier kinds; (4) Emergence of man from a common ancestor with apes; (5) Explanation of the earth's geology and the evolutionary sequence by uniformitarianism; and (6) An inception several billion years ago of the earth and somewhat later of life. <br> ***Arkansas Act 590 |

*Table 1 contd. ...*

*...Table 1 contd.*

| Year | Law |
|---|---|
| 2001 | **Santorum Amendment**\*\*\*\* (2001):<br>It is in the sense of the Senate that (1) good science should prepare students to distinguish the data or testable theories of science from philosophical or religious claims that are made in the name of science; and (2) where biological evolution is taught, the curriculum should help students understand why this subject generates so much continuing controversy and should prepare the students to be informed participants in public discussions regarding the subject.<br>\*\*\*\* Conference Report for the No Child Left Behind Act of 2001 |

we see a collection such laws and how the wording of these laws has developed and changed over time.

While the wording of each of these laws or legislation propositions varies greatly—the general meaning and purpose remains unmistakably clear. All of these legislative passages are designed to prevent students from learning about the theory of evolution, either by prohibiting the teaching of evolution in schools or by confusing students about the legitimacy of evolution by falsely claiming that scientists are still debating the validity of this major tenet of the biological sciences.

In this section, we will take an in-depth look at the cases that have taken place in the U.S. from 1925 to 2015 in order to understand the extent of the conflict. By looking at these legal cases, we will better understand how many conflicts start in the classroom or board room and expand into full-blown court cases. The driving force behind all of these cases is the deep emotional conviction that both sides have regarding education. While the creationists are most concerned with what is taught regarding human origins, the secularist are most concerned with keeping the wall between church and state erect. Above all, the extent of these legal cases highlights the severity of the 'us versus them' mentality that pervades the entire creationism | evolution conflict and here we must keep in mind that these cases are only the tip of the iceberg when it comes to the true societal division around the topic of evolution. In other words, these cases can be understood as the light cast by a candle into a dark cave. While these cases allow us to see the details of the conflict that made it into court, there is a myriad of other battles that did not find their way to the court houses due to lack of energy, time or funds (or because they settled out of court). The following pages contain a brief summary of each of the cases. A number of the original laws and other aspects of the cases can be found in the Appendix.

## *State of Tennessee v. John Thomas Scopes*[a]

The *Butler Act* (Table 1) was passed in Tennessee in 1925 making it unlawful "to teach any theory that denies the story of divine creation as taught by the Bible and to teach instead that man was descended from a lower order of animals." Soon after the bill was passed, the American Civil Liberties Union (ACLU) announced that they were looking for someone to challenge the *Butler Act* (Humes 2007). John Scopes, a biology teacher in Tennessee, volunteered to incriminate himself. The case was heard as a criminal case and the judge decided in favor of the prosecution, which was led by William Jennings Bryan. John Scopes was fined $100 for violating the law. According to historian Edward J. Larson, the ACLU was unconcerned with winning the case in Tennessee, since the ultimate goal was to appeal to the U.S. Supreme Court so that such anti-evolution statutes could be struck down at a national level (Larson 1997). Yet all did not go as planned.

The loss in the criminal court was appealed and while the Tennessee Supreme Court upheld the guilty verdict, they did overturn the $100 fine on a technicality. The decision to reverse to fine prevented the ACLU from being able to bring the case to the U.S. Supreme Court to test the constitutionality of the statute and thus the anti-evolution statute stayed on the books until the late 1960s.

This was the first trial in American history to feature live radio broadcasts of the proceedings and thus drew a great deal of attention to the conflict and controversy surrounding the teaching of evolution and illustrated the polarized nature of the American population—one side rallying for the advancement of science and one trying to protect their idea of an inerrant Bible. Although the case did not change any legal frameworks, it did have economic repercussions, in that publishers began to avoid or diminish the topic of evolution in subsequent years as they did not want to risk the chance of being boycotted/banned (Matzke 2010).

## *Susan Epperson et al. v. State of Arkansas*[b]

The Epperson case is historically linked to the Scopes trial as the central disagreement in the case revolved around the constitutionality of an Arkansas anti-evolution statute from the 1920s that had been modeled after Tennessee's *Butler Act* and was passed in Arkansas a couple years after the Scopes ruling verified that such legislation was in fact 'safe' (Numbers 1998). Moreover, this case was also orchestrated by a pro-science agency that actively sought out a teacher to violate the state statute with the aim of having it struck down (Cartwright 2004). In the 1960s the Arkansas Education Association approached Susan Epperson to participate in their attempt and Epperson agreed. According to Epperson, they chose her because she was and 'all-Arkansas' girl—white, southern and Christian (Cartwright 2004).

The case was heard by the State Chancery Court, which declared that the statute violated the 1st and 14th Amendment. The decision was reversed by the State Supreme Court stating that it is within the state's power to specify public school curriculum. Ultimately, the case was reheard by the U.S. Supreme Court, which upheld the original decision of the Arkansas State Court declaring that the Arkansas statute was in fact unconstitutional as it violated the Establishment Clause of the First Amendment.

This ruling by the Supreme Court brought an end to the 'Scopes Era' by making it unconstitutional for any state to have such anti-evolution legislation on its books and it established a national precedent that state curriculum could not "be tailored to the principles or prohibitions of any religious sect or dogma".

## *Joseph C. Daniel et al. v. Hugh Waters et al. & Harold Steele et al. v. Hugh Waters*[c]

In *Daniel v. Waters*, Tennessee biology teachers, parents and the National Association of Biology Teachers sued the Tennessee state textbook commission, including the chairman Hugh Waters claiming that the *Genesis Act* (Table 1) violated the First Amendment of constitution. At the same time Harold Steele and two other members of Americans United for Separation of Church and State also filed a suit, claiming that the Tennessee statute was in violation of the state and federal constitution. The Chancery Court ruled in 1974 in *Steele v. Waters* in favor of the plaintiffs, stating that the statute was in fact in violation of the First and Fourteenth Amendments. The state of Tennessee appealed this ruling to the Supreme Court of Tennessee. In 1975, the U.S. Court of Appeals, Sixth Circuit decided in *Daniel v. Waters* that the Tennessee statute was "patently unconstitutional". In 1975, the Supreme Court of Tennessee ruled in favor of the plaintiffs in *Steele v. Waters* and also concurred with the *Daniel v. Waters* decision in its brief opinion.

This case clearly showed that it is unconstitutional for states to dictate the content of textbooks based on religious premises, i.e., a state cannot prohibit the use of textbooks where evolution is portrayed as one of many theories instead of scientific fact and furthermore it is unconstitutional to require the equal-treatment of evolution and the Genesis story. Yet the limited jurisdiction in the case did not prevent creationists from continuing to champion similar equal-time statutes in other states.

## *John Hendren et al. v. Glenden Campbell et al.*[d]

This case involved the Indiana's Textbook Commission's approval of *Biology: A Search for Order in Complexity*, which was integrated into public school classrooms to teach biology courses. This 'creation science' textbook was published by the Creation Research Society in 1970 and was promoted through the Institute for Creation Research (see Appendix I for full product description). The suit was led by John Hendren, a ninth-grade student, his father, Jon Hendren as well as E. Thomas Marsh, another school parent. They sued the members of the commission, Campbell et al. as individuals and in their capacity of commission members, claiming that the use of a book that promotes creationist concepts in public schools is a violation of the students' constitutional rights. The Marion Superior Court ruled in favor of the plaintiffs, stating that it was clearly unconstitutional to use a creationist or 'creation science' textbook in public schools. As Judge Dugan said, "The question is whether a text obviously designed to present only the view of Biblical Creationism in a favorable light is constitutionally acceptable in the public schools of Indiana. Two hundred years of constitutional government demand that the answer be no."

## *Segraves et al. v. State of California et al.*[e]

Kelly Segraves, as head of the Creation-Science Research Center, sued the State of California Board of Education on behalf of his three school children arguing that his family's right to free exercise of religion was violated by the discussion of evolution in their public school. A number of republican politicians as well as a member of the state board joined Segraves in his suit against the State of California, the Board of Education and two democratic politicians. The judge ruled in favor of the defendants, stating that the teaching of evolution cannot be considered the establishment of religion and that such infringements are prevented by State Board of Education's 1972 anti-dogmatism policy (see Appendix I for full text). This ruling established that the teaching of evolution in science classes is constitutionally legitimate and not an infringement of the students' right to exercise their own religious beliefs.

## Reverend William McLean et al. v. Arkansas Board of Education et al.[f]

Reverend William McLean (United Methodist Church) along with other clergy men, parents of children attending Arkansas public schools and a biology teacher challenged the constitutionality of the *Balanced Treatment for Creation-Science and Evolution Science Act* (Table 1), which was enacted in Arkansas in 1981 and required that if evolution was taught in a classroom, then an equal amount of time must be dedicated to the teaching of 'creation science'. Despite the defendants attempt to show the 'scientific validity' of creation science, the court declared that the Act violated the Establishment Clause of the First Amendment.

The ruling was not binding outside the jurisdiction of the court but the detailed decision, by Judge William Overton, included a specific definition of science and ruled that creation science is religion and not science. Although the ruling was not binding outside of the jurisdiction of the court, the decision provided the precedent and foundation for cases involving the teaching of creation science—most notably *Edwards v. Aguillard* (Forrest and Gross 2007). This case also highlighted the fact that religious leaders were as equally concerned with the teaching of creation science as science educators.

## Edwin W. Edwards v. Don Aguillard (Appeal of the ruling from Aguillard v. Treen)[g]

Louisiana's *Creationism Act* was passed in 1981 and prohibited the teaching of the theory of evolution in public elementary and secondary schools unless it was accompanied by instruction of 'creation science'. The constitutionality of this Act was challenged by a group led by biology teacher Don Aguillard and twenty-six other organizations and individuals including parents, teachers, and religious leaders. The plaintiffs claimed that it is unconstitutional for Creation Science to be taught in an equal and balanced manner with evolution in public school classrooms. The district court ruled in favor of Aguillard and the other plaintiffs, stating that the Act was in fact in violation of the Constitution. The defendants, including new Governor Edwin W. Edwards, appealed the decision. The District Court granted summary judgment in favor of the Respondents, which was affirmed by the Fifth Circuit Court of Appeals. The U.S. Supreme Court then ruled that the Act violated the Establishment Clause of the First Amendment. The decision was based on the three-pronged Lemon Test[5] and the detailed ruling from *McLean v. Arkansas*.

The ruling affected all public schools in the United States by making it illegal to teach creationism or creation science in public schools because it is a clear attempt at the advancement of a particular religion. While this ruling led to the end of 'balanced treatment' legislation, it also spawned a new type of creationism: intelligent design which was focused more on finding scientific evidence for the presence of an intelligent designer instead of trying to find data to prove the Genesis account.

## Ray Webster et al. v. New Lenox School District #122 et al.[h]

Social studies teacher, Ray Webster, and a student in his class, Mathew Dunne, sued the New Lenox School District and the superintendent because he believed that his 1st and 14th Amendment rights had been violated when Webster was prohibited from teaching his creationist beliefs in the classroom. Dunne likewise argued that he had a right as a student to hear about creationism or creation science to balance the teaching of pro-evolution statements. The U.S. District Court for the Northern District of Illinois court found in favor of the school district and superintendent on the grounds that a teacher does not have the right to advocate religion in the classroom. Moreover, the court ruled that Dunne's desire to learn about creation science was outweighed by the district's obligation to avoid violating the Establishment Clause and thus upholding the other students' first amendment rights.

This case highlighted the fact that teachers do not have an unlimited right to the freedom of speech and that they, too, are bound by the Establishment Clause and are unable to promote a particular religious view in their classrooms. While teachers do not forfeit the right to comment publicly on matters of public importance simply because they accept a public-school teaching position, a teacher is a representative of the school district when he/she teaches in the classroom and thus bound by the same Constitutional regulations as the school district.

## John E. Peloza v. Capistrano Unified School District[i]

Biology teacher John E. Peloza sued the Capistrano Unified School District for forcing him to teach evolution as scientific fact and for prohibiting him from discussing his religious beliefs with students during instructional time. Peloza claimed that these policies violated his freedom of speech, freedom of religion, and rights to due process and equal protection (1st, 5th and 14th Amendment). As part of his case, he argued that 'evolutionism' was part of the 'religion of secular humanism'. The court ruled against Peloza on all accounts in 1992. In its ruling,

---

[5]    The 'Lemon Test' was articulated by the Supreme Court in 1971 in the case of *Lemon v. Kurtzman* and contains three prongs: purpose, effect and entanglement. Purpose means that the government (or here public school) action must have a secular purpose. Effect means that the effect of the policy may neither support nor inhibit religion. Entanglement means that the result of the action may not be an excessive entanglement of government with religion.

the U.S. District Court Central District of California used the 'Balancing Test'[6] and determined that "the interests and concerns of the school district overrule the plaintiff's claimed right to free speech." The interest of the school district was defined as "maintaining its secular purpose of educating high school students" (Wallis 2005). Peloza appealed the case but the U.S. Court of Appeals upheld the dismissal of Peloza's claim in 1994.

## *Herb Freiler et al. v. Tangipahoa Parish Board of Education et al.*[j]

In 1994, the Tangipahoa Parish school board voted to have a disclaimer (see Appendix I for full disclaimer) read before the topic of evolution was taught. Three years later, a group of parents sued the board of education, the members of the school board and the school superintendent claiming that reading of such a disclaimer was a promotion of religious thought and thus a violation of the Establishment Clause of the First Amendment. The U.S. District Court for Eastern District of Louisiana found in favor of the parents in 1997. The school board appealed the ruling to the U.S. Court of Appeals, where the decision was upheld in 1999. The U.S. Supreme Court declined to hear the case, thus allowing the lower court ruling to stand.

## *Rodney LeVake v. Independent School District 656 et al.*[k]

High school biology teacher Rodney LeVake refused to teach evolution unless he was able to teach "the difficulties and inconsistencies of the theory". He was therefore reassigned to a ninth-grade general science course. In 1999, LeVake sued the school district, the superintendent, the school principal and the curriculum director in order to recover his original position. LeVake alleged that the reassignment violated his right to free exercise of religion, free speech, due process, and academic freedom (1st and 14th amendment rights). He also claimed that the district's teaching assignment policy was illegal according to the United States and Minnesota Constitutions. In 2000, the Minnesota State District Court ruled against LeVake on all accounts in favor of the school district's curriculum policy.

## *Tammy Kitzmiller et al. v. Dover Area School District et al.*[l]

In 2005, a group of parents sued the Dover Area School District and the Board of Directors when the school board required that a statement be read aloud in ninth-grade science classes when evolution was taught. The statement presented Intelligent Design as 'an explanation of the origin of life that differs from Darwin's view' and referred students to copies of the creationist textbook, *Of Pandas and People*, available at the school library. The plaintiffs claimed that the reading of the disclaimer

---

[6]   The 'Balancing Test' can be used to determine whether or not a public employee's speech is protected by the First Amendment. The test is "a balance between the interests of the employee, as a citizen, in commenting upon matters of public concern and the interest of the state, an employer, in promoting the efficiency of the public services it performs through its employees" as first defined in *Picerking v. Board of Education* in 1968. The test was first used by the U.S. Supreme Court in Rankin v. McPherson in 1987.

was a violation of the students' First Amendment rights. The ACLU represented the group of parents. In making his ruling, Judge Jones applied the Lemon Test and the Endorsement Test[7] and ruled that the teaching of intelligent design in public school biology classes is a violation of the Establishment Clause of the First Amendment. They found that intelligent design is not science and "cannot uncouple itself from its Creationist, and thus religious, antecedents."

## Kenneth Hurst et al. v. Steve Newman et al.[m]

The Americans United for Separation of Church and State filed suit against the members of the El Tejon Unified School District School Board, the superintendent, the school principal of Frazier Mountain High School and a teacher on behalf of eleven parents for introducing a course entitled 'Philosophy of Intelligent Design'. The course specifically promoted creationism and intelligent design while undermining evolution as seen in the course description:

> *This class will take a close look at evolution as a theory and will discuss the scientific, biological, and Biblical aspects that suggest why Darwin's philosophy is not rock solid. This class will discuss Intelligent Design as an alternative response to evolution. Topics that will be covered are the age of the earth, a world wide flood, dinosaurs, pre-human fossils, dating methods, DNA, radioisotopes, and geological evidence. Physical and chemical evidence will be presented suggesting the earth is thousands of years old, not billions. The class will include lecture discussions, guest speakers, and videos. The class grade will be based on a position paper in which students will support or refute the theory of evolution.*[8]

The plaintiffs claimed that offering such a course in a public school is a violation of the students' first amendment rights. The case was dismissed with prejudice, and the El Tejon School District settled the lawsuit in 2006 by agreeing to cancel the course and to never offer another course "entitled 'Philosophy of Design' or 'Philosophy of Intelligent Design' or any other course that promotes or endorses Creationism, Creation Science, or Intelligent Design."

## Jeffrey Michael Selman et al. v. Cobb County Cobb County School District, Cobb County Board of Education[n]

The Cobb County School District began the process of textbook adoption in 2001. A group of parents was concerned about the new emphasis on the teaching of evolution

---

[7]    The 'Endorsement Test' is used to determine what message a certain governmental policy or enactment is trying to convey. The Endorsement Test was adopted by the Supreme Court in 1989 in *County of Allegheny v. ACLU* which involved various holiday displays near a courthouse in Pennsylvania. The Endorsement Test asks whether a "reasonable observer" would feel that the governmental action (or specifically in evolution cases, a school board's policy) sends a "message to non-adherents that they are outsiders, not full members of the political community, and an accompanying message to adherents that they are insiders, favored members of the political community" (Lynch v. Donnelly, U.S. 465: 668, 687. 1984).

[8]    https://ncse.com/creationism/legal/hurst-v-newman-2006 (Accessed 27 October 2017).

and complained that the teachers should be teaching creationism. In 2002, the school board thus decided to insert a disclaimer into the textbooks to accommodate the religious views of the parents. In 2004, a different group of parents sued the Cobb County School District and Cobb County Board of Education claiming that a disclaimer that warns students about evolution is a promotion of a particular religious view point and is therefore in violation of the Constitution. In 2005, the U.S. District Court Northern District of Georgia found that the use of the disclaimer was in violation of the Georgia State Constitution[9] because it had been paid for using public funds. The court also found that the use of disclaimers violated the Establishment Clause of the First Amendment. The case was appealed and settled in 2006.

## *Larry Caldwell v. Roseville Joint Union High School District*[o]

Larry Caldwell sued the Roseville School District, the Board of Trustees, the superintendent, the assistant superintendent, the deputy superintendent and the principal of Granite Bay High School in 2005. Caldwell claimed that his right of free speech and freedom to practice religion had been violated when the school district rejected his proposed "Quality Science Education Policy", which would require teachers to teach the "scientific strengths and weaknesses" of evolution. According to Caldwell's proposal "Because 'nothing in science or in any other field of knowledge shall be taught dogmatically' and 'scientific theories are constantly subject to testing, modification, and refutation as new evidence and new ideas emerge', teachers in the Roseville Joint Union High School District are expected to help students analyze the scientific strengths and weaknesses of existing scientific theories, including the theory of evolution." Caldwell's policy suggestion followed Caldwell's failed attempt at the prevention the use of the *Holt Biology Textbook* which in his opinion is not accurate, objective or current. The court dismissed all of Caldwell's claims in 2005 and again in 2007 denying that the school district had violated any Caldwell's constitutional rights.

## *Association of Christian Schools International (ACSI) et al. v. Roman Stearns et al.*[p]

The University of California has a policy of rejecting certain biology classes from Christian private schools due to the "inconsisten[cy] with the viewpoints and knowledge generally accepted in the scientific community." The Association of Christian Schools International, Calvary Chapel Christian School, and a group of parents filed a lawsuit against UC officials in 2005 claiming that the university's policy violated applicants' constitutional rights. The original lawsuit against the university officials was dismissed in 2006, but the judge allowed the case against the university system to continue. In 2008, the judge ruled that the policy was proper and constitutional. The plaintiffs appealed the case. The decision was upheld by the

---

[9]   Georgia State Constitution Article 1, paragraph 2, section 7: No money shall ever be taken from the public treasury, directly or indirectly, in aid of any church, sect, cult, or religious denomination or of any sectarian institution.

Ninth Circuit Court of Appeals in 2010 and U.S. Supreme Court declined to review the case. While the court did rule in favor of UC, there was considerable public support for the plaintiffs as illustrated by the submitted Amicie Curiae Brief:[10]

> *The proper resolution of this case is a matter of substantial concern to amici due to the impact it will likely have on religious education in California and across the country. Amici urge this Court to rule in Plaintiffs-Appellants' favor because the First Amendment prohibits the religious discrimination that is pervasive in the University of California's selective scrutiny of the curriculum of religiously affiliated private schools. California students should be considered for admission to a state university without regard to their religious worldviews.*

## C.F. v. Capistrano Unified School District[q]

Dr. James Corbett, a history teacher in the Capistrano Unified School District, allegedly referred to creationism as "superstitious nonsense" during one of this classes. One of his students, Chad Farnan, took offense at the statement and his parents decided to sue both Corbett and the school district in 2007, claiming that Corbett's statements violated their son's First Amendment rights as they were an "exhibition of hostility toward religion and endorsement of irreligion in a public-school classroom." The District Court for Central California used the Lemon Test to determine the constitutionality of Corbett's remark and found that his comment about creationism lacked a secular purpose and "constitutes improper disapproval of religion in violation of the Establishment Clause." Despite these findings, the court denied Farnan's request for an injunction against Corbett or the Capistrano Unified School District. The Federal Ninth Circuit Court of Appeals upheld Corbett's immunity and declined to rule on the constitutionality of his remarks stating that the issue was resolved 'on [the] basis [of qualified immunity] alone'.

## Christina Comer v. Robert Scott and Texas Education Agency (TEA)[1]

Christina Comer had worked as the Director of Science for the Curriculum Division at the Texas Education Agency for over 10 years when she was fired for failing to remain 'neutral' by sending an email to science educators about a lecture addressing creationism and evolution. In 2008 Comer filed suit against Robert Scott, the commissioner TEA claiming that the agency's 'neutrality policy' in regard to teaching creationism as science in public schools was in violation of the Establishment Clause as it promoted religion by treating creationism as a valid scientific theory. The lawsuit was dismissed in 2009 by the district court. Comer appealed the decision, but the decision of the lower court was upheld in 2010.

---

[10]    Amicie: American Center for Law and Justice, Catholic League for Religious and Civil Rights, Common Good Foundation and Church State Council of Seventh-day Adventists.

## *Jeanne Caldwell v. Roy Caldwell et al.*[s]

Jeanne Caldwell[11] filed suit against UC Berkeley professors, Roy Caldwell and David Lindberg, and Michael Piburn, Program Director for the National Science Association, claiming that the creation and content of their 'Understanding Evolution' website (http://evolution.berkeley.edu/) was in violation of the 1st and 14th Amendment because certain religious beliefs were allegedly endorsed while other religious beliefs were censored. The defense responded by pointing out that the purpose and effect of the website was not to promote religion but was designed to help K-12 teachers teach evolution. The specific pages mentioned by the plaintiff were designed to debunk the misconception that evolutionary theory and religion are incompatible. In 2006, the U.S. District Court for the Northern District of California dismissed the case without ruling on the merit of the constitutionality of the claim because the plaintiff was unable to prove her taxpayer status or a concrete injury.

## *Doe v. Mount Vernon Board of Education et al.*[t]

In 2010, an anonymous family filed a lawsuit against the Board of Education of the Mount Vernon City School District, the district's superintendent, the principal of Mount Vernon City School, and against an eighth-grade science teacher, John Freshwater. They claimed that Freshwater had violated the First Amendment by attacking evolution, displaying religious objects, leading a prayer session and teaching intelligent design. They also filed suit over physical injury caused by Freshwater, who branded the Doe's son with the sign of the cross. There was a court settlement in favor of the Doe family. Freshwater was fired and has since filed a lawsuit against the Mount Vernon School Board for unfair dismissal.

## *Institute for Creation Research Graduate School (ICR) v. Raymund Paredes et al.*[u]

In 2005, the Institute for Creation Research (ICR) created a new online master's degree in science education designed to instruct teachers on the scientific evidence for creation and how to teach those truths in classrooms. The Texas Higher Education Coordinating Board (THECB) denied the ICR the authority to issue such a master's degree in science education. The ICR thus filed suit against the CEO of the THECB and other officers of the board claiming that denying the ICR the authority to issue a master's degree in science education was a violation to the ICR's 1st and 14th amendment rights. The U.S. District Court for Northern District of Texas ruled in favor of the defendants after the plaintiffs were unable to provide material evidence.

## *American Freedom Alliance v. California Science Center*[v]

The American Freedom Alliance (AFA) sued the California Science Center (CSC) in 2011 after the CSC canceled a screening of the film *Darwin's Dilemma—The*

---

[11]   This was the second case involving the Caldwell family (see *Caldwell v. Roseville*).

*Mystery of the Cambrian Explosion*. The film promotes intelligent design and the AFA claimed that the cancellation was based on the content of the film and therefore a violation the First Amendment. The Superior Court for the State of California dismissed the case and the parties settled out of court. Neither party accepted any fault or liability. Yet AFA claimed that it was a free speech case win for the ID movement since the CSC paid $110,000 to the AFA.[12] It was later discovered that the Discovery Institute was working in coordination with the AFA in order to increase the controversy and provoke a cancellation.[13]

## *Pamela Hensley v. Johnston County Board of Education et al.*[w]

In 2004, Pamela Hensley was teaching her 8th grade science class about evolution when a lively debate took place. In 2005, parents of one of the students complained that Hensley had been rude to their daughter by saying that the Bible was not to be read literally and allegedly punished her for her religious view by giving her a poorer grade. Hensley was reprimanded by the principal. In 2005 the same father met with the school board and demanded that Hensley publicly admit that she had demonstrated "unconstitutional hostility against the beliefs of the Christian students in the classroom by questioning the literal content of the Bible and by teaching her theological position that the Bible contains errors". The father also requested that Hensley be transferred out of the North Johnston school district and be assigned to teach a subject other than science. Hensley was transferred to a remedial arts school a couple of weeks later. In 2007 Hensley filed suit against the school board claiming the demotion and transfer violated her right to freedom to speech. In 2010 the U.S. District Court for the Eastern District of North Carolina granted the defendants their Movement to Dismiss Hensley's claims regarding the Constitutional violations. This case highlighted how careful teachers must be when discussing evolution in a classroom and that even by stating that the Bible should not be read literally may put a teacher's job at risk.

## *Scott and Sharon Lane v. Sabine Parish School Board et al.*[x]

C.C. Lane, a Thai Buddhist, enrolled in Negreet in the 6th grade in 2014 where he was quickly harassed and prolystetised by his science teacher, Rita Roark. Roark continually promoted her Christian beliefs. For example, she included fill in the blank statements such as "Isn't it amazing what the _____ has made!!!!" as a compulsory science exam questions requiring students to fill in the word "Lord". The Lane family complained to the superintendent in 2014, claiming that continual promotion and teaching of Christian belief in public schools while simultaneously punishing students for not participating is in violation of that student's constitutional rights. The superintendent informed them that they were in the Bible Belt and would

---

[12]  *California Science Center Pays $110,000 to Settle Intelligent Design Discrimination Lawsuit.* Evolution News and Views. August 29, 2011. www.evolutionnews.org. (Accessed March 15, 2015).

[13]  California Science Center Foundation's Statement Regarding Resolution of Legal Dispute with AFA. PR News Wire. August 29, 2011. www.prnewswire.com (Accessed March 9, 2015).

have to accept being prolystetised by teachers such as Roark. The Lane family was further told that Buddhism was stupid, and that C.C. should conform to Christianity or go to another school where there are more Asians. The plaintiffs filed suit against Sabine Parish School Board, the superintendent, the school principal and Rita Roark. The Lanes were offered a Decree of Consent, which they accepted stipulating that the district-wide promotion of religion would be in violation of the Establishment Clause if proved. The Board was also required to bus Lane to another school for the remainder of his education.

## Summary

This very brief overview of the many legal cases involving creationism and evolution illustrates many important points. First of all, from a legal standpoint, we see that the foresight of the founding fathers in the creation of the Constitution, the Bill of Rights and the separation of church and state has offered judges in the 20th and 21st century a clear foundation upon which they can make their rulings regarding the illegitimacy of teaching creationism in public-school science classrooms. This enlightened foresight made it possible for almost all of the case rulings to be made in favor of the pro-secular education.

At a personal level, it is important to notice how many of these cases involve individuals (such as parents and teachers) challenging the policies created by larger organizations, whether it be state legislation, university policies or school board decisions. The active involvement and dedication seen in all of these individuals is an expression of the deep-seeded belief systems and emotions that act as the bedrock of the evolution-creationism conundrum. This is true for both the pro-creationist individuals as well as the pro-secular-education individuals. Each group is convinced of the correctness of their own viewpoint and both sides are equally convinced that the court will rule in their favor.

This ardent effort highlights the intense emotions involved in the cases and the fact that legal action was pursued points to the fact that there is a great deal of *negative* emotions, i.e., fear, mistrust and anger, surrounding the topic of evolution education. It is obvious that parents and other members of society are concerned about the relationship between their faith and the effect that evolution could have on that faith. In other words, there is a general fear that learning about evolution could cause students to accept or even believe in evolution, which would thus lead a student to question the literal interpretation of the Bible. This fear highlights their passion for upholding a literalist interpretation of the Bible and their perceived necessity of such an interpretation for the 'correct' order of society.

At a broader societal level, it is important to point out that while many publications, studies and statistics highlight the prevalence of creationism in the United States, this data should not overshadow the efforts and dedication of evolution-accepting, science-loving Americans who are equally willing to go to court to defend their belief in secular education. For every creationist effort to introduce non-scientific material into the science classroom, there is an equally dedicated effort made to preserve secular science education. The prevalence of legal cases underscores the degree of conflict surrounding this topic and we see that both sides are engaged in a clear 'us vs. them' mentality. There is a true dichotomy present in this situation that reflects the increasing division among the American public. Both sides feel like they are in the right and that the other party is in the wrong. Instead of engaging in conversations, the conflicting parties are engaged in debates and legal disputes. It is obvious that the roots of this conflict run deep and are intrinsically connected to our personal and cultural identity as well as our worldviews.

The profound involvement of cultural identity and emotions in this situation means that there are no easy or superficial solutions. Thus, we will have to develop a much better understanding not only of the creationist belief system but of the role that these beliefs play in the lives of a creationist and how these beliefs directly hinder a person's ability to comprehend the true richness of natural history.

---

[a]　Year: 1925; State: Tennessee; Courts: Criminal Court of Tennessee to Tennessee Supreme Court; Appeal Citation: 154 Tenn. 105 (1925), 289 S.W. 363 (1927).

[b]　Year: 1968; State: Arkansas; Court: Arkansas State Court (Chancery) to Arkansas State Supreme Court to US Supreme Court; Appeal Citation: Epperson v Arkansas, 393 U.S. 97 (1968).

[c]　Year(s): 1973–1975; State: Tennessee; Courts: Chancery Court, U.S. Court of Appeals, Sixth District and the Supreme Court of Tennessee; Citation: Daniel v Waters, 515 F.2d 485 (6th Cir. 1975).

[d]　Year: 1977; State: Indiana; Court: Marion Superior Court, NO. 5 (Marion county, Indiana); Citation: Hendren v Campbell, Superior Court No. 5, Marion County, Indiana, 14 April 1977.

[e]　Year: 1981; State: California; Court: Superior Court of California; Citation: Segraves v. California, No. 278978 (Super. Ct. Sacramento County 1981).

[f]　Year: 1982; State: Arkansas; Court: U.S. District Court for the Eastern Districts of Arkansas; Citation: McLean v Arkansas Board of Education, 529 F. Supp. 1255, E.D Ark. (1982).

[g]　Year: 1987; State: Louisiana; Court: US Court of Appeals, 5th Circuit to the U.S. Supreme Court; Citation: Edwards v Aguillard, 482 U.S. 578 (1987).

[h]　Year: 1989; State: Illinois; Court: US District Court for the Northern District of Illinois; Citation: Webster v New Lenox School District #122, 917 F.2d 1004 (7th Cir. 1990).

[i]　Year: 1994; State: California; Court: US District Court Central District of California to U.S. Court of Appeals, 9th Circuit; Citation: John E Peloza v. Capistrano Unified School District, 37 F.3d 517 (9th Cir. 1994).

[j]　Years: 1997/2000; State: Louisiana; Court: US District Court for Eastern District of Louisiana (1997) to US Court of Appeals for the Fifth Circuit (2000); Citation: Freiler v. Tangipahoa Parish Board of Education, 185 F.3d 337 (5th Cir. 1999).

[k]　Year: 2000; State: Minnesota; Court: Minnesota State District Court; Citation: LeVake v Independent School District No. 656, 625 N.W.2d 502 (MN Crt Appl. 2000), cert. denied, 534 U.S. 1081 (2002).

[l]　Year: 2005; State: Pennsylvania; Court: U.S. District Court for the Middle District of Pennsylvania; Citation: Kitzmiller v. Dover, 400 F. Supp. 2d 707 (M.D. Pa. 2005).

[m]　Year: 2006; State: California; Court: U.S. District Court, Eastern District of California; Citation: 06-036 - Kenneth Hurst, Et Al. v. Steve Newman, et al.

[n]　Year: 2006; State: Georgia; Court: U.S. District Court Northern District of Georgia to U.S. Court of Appeals for the 11th Circuit; Citation: 449 F.3d 1320 (11th Cir. 2006).

[o]　Year: 2005/2007; State: California; Court: U.S. District Court Eastern District of California; Citation: 05-061 - Caldwell v. Roseville Joint Union High School District.

[p]　Year: 2008; State: California; Court: U.S. District Court of Central California to Ninth Circuit Court of Appeals; Citation: No. CV 05-6242 SJO (MANx); Case 2:05-cv-06242-SJO-MAN.

[q]　Years: 2007–2011; State: California; Courts: U.S. District Court for Central California to Federal Ninth Circuit Court of Appeals; Citation: 647 F. Supp. 2d 1187 (C.D. Cal. 2009).

[r]　Years: 2009/2010; State: Texas; Court: U.S. District Court, Western District of Texas to U.S. Court of Appeals, 5th Circuit; Citation: Comer v. Scott, 610 F. 3d 929 - Court of Appeals, 5th Circuit 2010.

s   Year: 2006; State: California; Court: US District Court for the Northern District of California, 9th Circuit Court of Appeals; Citation: 3:05-CV-04166-PJH.

t   Year: 2010; State: Ohio; Court: U.S. District Court for the Southern District of Ohio, Eastern Division; Citation: 2:08-cv-00575-GLF-NMK.

u   Year: 2010; State: Texas; Court: U.S. District Court for Northern District of Texas, U.S. District Court for the Western District of Texas; Case No.: A-09-CA-382-SS.

v   Year: 2011; State: California; Court: Superior Court for the State of California, County of L.A. - Central District; Case No.: BC 423867.

w   Year: 2010; State: North Carolina; Court: U.S. District Court for the Eastern District of North Carolina; Case No.: 5:07-CV-231.

x   Year: 2014; State: Louisiana; Court: U.S. District Court Western District of Louisiana; Case No.: 5:14-cv-00100-EEF-KLH.

# Chapter 3
# Morphology of Misconceptions

While we now have a good sense of the historical, societal, cultural, religious and legal complexity of creationism, the main goal of this book is to understand creationist thinking and the perpetual clinging to literalist beliefs at a cognitive level in order to find a means by which students and the general public can become more receptive to science. For that reason, this chapter will dissect creationist thought in order to understand the nature of creationism and to illustrate the variation of creationist beliefs. We will also look at how these beliefs lead to varying degrees of misconceptions based on the degree of literalism and the degree to which a creationist internalizes these beliefs, thereby affecting their ability to process and comprehend the validity of evolutionary science.

This analysis of the belief systems and the categorization of the misconceptions allows us to better understand how creationist beliefs differ from other educational misconceptions in both their scope and tenacity. It should be noted that misconceptions or at least preconceptions are a natural phenomenon. There is not a student in this country (or world) who enters the classroom as a blank slate. Even at the youngest of ages, children have acquired a certain understanding of the world around them. Yet, due to their inability to understand the complexity of the world around them, they often bridge the gaps in their understanding with simpler explanations.

There are also many preconceptions and even misconceptions about evolution that do not necessarily pose a threat to science education or a person's potential scientific literacy. I will outline a couple examples of these types of misconceptions that arise due to lack of knowledge or an incorrect assimilation of knowledge.

Many misconceptions about the natural world are the product of simple assumptions that are used in place of exact knowledge about a certain subject. A great example of this is seen in how children (and adults) construct phylogenic relationships based on superficial observations because previous experience has taught them that animals that look alike are often closely related, e.g., foxes and domestic dogs. In this way, many could assume that sharks and dolphins are related or that hedgehogs and porcupines are related. They would not *instinctively* assume that dolphins are more closely related to cows than they are to sharks or that porcupines are more closely related to hamsters than to hedgehogs. In order to understand why a dolphin is considered a mammal and not a fish, a child or student must first understand a great number of things about animal physiology and zoological classification. In the absence of this knowledge, it is possible and likely that a child would group all flying animals together into one group and all swimming animals into another. These types of misconceptions are often easily remedied once the child receives the previously lacking information. Rectifying these types of misconceptions can even be associated with positive emotional experiences such as exhilaration in the acquisition of new (shocking) information such as the fact that whales and dolphins still have remnants of pelvic bones and that both the porcupine and the hamster have incisors that grow continuously throughout their lifetime.

Other misunderstandings about the natural world are based on a false assimilation of knowledge. When my daughter was around four, we went to the Phyletic Museum in Jena, Germany which was established by Ernst Haeckel as an educational institution to visualize evolutionary theory. It was during this trip to the museum that we first talked about evolution in detail as we looked at the exhibition on human evolution. Not soon thereafter we visited the local zoo, where she sat in front of the spider monkey enclosure for close to an hour (Figure 7). While I do love observing monkeys, I did try to persuade her that it was time to move on to another part of the zoo, but she refused. When I asked her why—she simply responded that she was waiting for their tails to fall off and turn into humans. After my initial response—"What?!"—I realized that she had seen a number of exhibitions that day in the museum that were all set up in the same left-to-right transformation processes: caterpillars to butterflies, tadpoles to frogs, fetuses into babies and yes prehistoric apes into hominids. Apparently in the mind of a child—these were all similar processes, and all occurred relatively quickly. We had already raised a number of tadpoles at home and during the later stages of development, the changes (e.g., loss of a tail) do in fact appear to happen overnight.

**Figure 7:** Zoey waiting for the monkeys' tails to fall off at the Erfurt Zoo—an example of a natural (immature) misconception of evolution that attributes the characteristics of metamorphosis to evolution.

So, I made my second attempt at explaining and clarified that evolution does not happen over the course of a summer afternoon or even within one lifetime but instead over many generations—even millions of *generations*. She seemed to accept this explanation and we were able to move onto the next group of animals. Later that week, though, I heard her 'explaining' to her father that all human babies were once a type of monkey called 'feet-uses' before they are born. It was at this point that I realized that I needed help and bought a number of books on evolution for kids (more information regarding those publications is available in the last chapter, Chapter 8: Understanding Common Descent).

Zoey's completely false understanding of evolution was due to a false assimilation of knowledge, combining newly acquired information about human evolution with what she understood about metamorphosis and prenatal development. Admittedly, it may have also been caused by my poor explanation or my failure to realize that she was too young to truly grasp evolutionary theory. Regardless of the cause, these types of misconceptions are a natural part of learning and are often self-correcting in nature as children mature.

It could therefore be argued that the majority of false beliefs or misconceptions that children bring into the classroom are innocuous because they are often easily rectified once the child has received the correct information and has reached the age at which they can comprehend that information. Yet there are other types of misconceptions that are not innocuous and instead pose a serious threat to scientific literacy. This is seen most clearly in the misconceptions that lead individuals to reject evolution or deny data concerning anthropogenic climate change. While I briefly discuss climate science denial at the end of this book, here we will continue to focus on the rejection of evolution.

Creationist beliefs represent a major hinderance to scientific literacy because they are not the product of a simple misunderstandings, a lack of information, a false assimilation of information or immaturity. Creationist beliefs are instead deeply entrenched in a person's sense of purpose, their identity and their understanding of the world around them. However, not all forms of creationist thought cause the same degree of scientific illiteracy, since there is a great deal of variety within this belief system. Creationist thought can thus be best understood as a spectrum, where certain groups adhere to more literalist interpretations of the Bible than others. In order to avoid false generalizations, I will first offer a brief overview of creationist thought variations and then later in this chapter I will introduce a categorization of misconceptions to better explain the varying degrees of tenacity of misconceptions and their resulting potential to disrupt a person's ability to receive, accept and understand information on evolution.

## Creationist Thought

*"To claim that evolution is 'just a theory' is to reveal both a profound ignorance of modern biological knowledge and a deep misunderstanding of the basic nature of science."*                                                                    T. Ryan Gregory

While creationists and their beliefs are often discussed as if they are homogeneous and static, it is crucial to understand that creationist thought is actually rather diverse and thus best understood as a spectrum. On one side of the spectrum, we can find a person of faith who firmly believes in a personal God but is nevertheless comfortable with the fact that God used evolution as a tool to create life—such persons of faith are often referred to as supporters of theistic evolution. On the other side of the spectrum we have a person with an equally strong degree of faith in God but instead of accepting the allegorical nature of the Bible, they are convinced of the necessity of interpreting the Bible literally and are thus resistant to or actively opposed to teachings on evolution. It is this component of literalist thinking that is the true hallmark of fundamentalism and thus the major differentiation between creationists and other mainline Protestants. Yet, even among the creationists there is a large degree of variation in the scope of their literal beliefs. Eugenie Scott, former director of the National Center for Science Education, first proposed the idea of a creationist continuum in 2009 and created the following diagram to illustrate her concept of this spectrum (Figure 8).

While I am in total agreement with Scott regarding the spectrum of creationist thought, I have altered the diagram slightly to highlight the differences in the degree of literalist thinking. Moreover, I have changed the division lines from Old Earth and New Earth to a focus on the New Testament versus the Old Testament since the focus on the New Testament allows for a more allegorical understanding of the Genesis account (Figure 9).

The most literal of all creationists are those who still believe in a flat earth as it was described in various books of the Old Testament, perhaps most clearly in Isaiah 40:22 "He sits enthroned above the circle of the earth, and its people are like grasshoppers. He stretches out the heavens like a canopy, and spreads them out like a tent to live in." Some flat-earth creationists also refer to the passages in the New Testament where Jesus *ascended up into Heaven*—apparently also a clear description of a flat Earth.

Another group of very literalist creationists are geocentric creationists, who reject the majority of all modern physics, astronomy and biology. While geocentric creationists take one step forward from the flat-earth creationists and accept that the Earth is a globe, they still claim that the Earth is the center of the solar system. Again, these claims are based on a strict literal interpretation of particular passages from

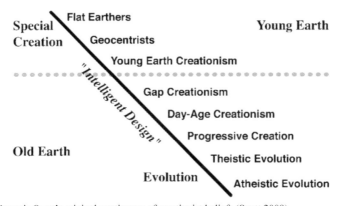

**Figure 8:** Eugenie Scott's original continuum of creationist beliefs (Scott 2009).

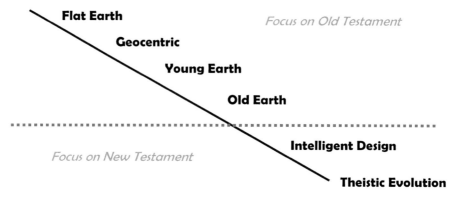

**Figure 9:** Proposed modification of the creationist spectrum with emphasis on the difference between literalist and allegorical understandings of Genesis.

the Bible. The most important passage for geocentric thought is Joshua 10: 12–14, where the sun stands still (i.e., stops orbiting around the earth). This passage is also interesting as it is Joshua who calls for the sun to stop and the Lord listened to him. Thus, this passage also highlights God's interest in man's affairs and the personal relationship possible with God.[14]

Taking it another step forward, the young-earth creationists (YECs) accept that the Earth is a globe and accept the concept of heliocentrism (i.e., that the sun is the center of the solar system), but they deny all scientific data regarding the age of the Earth. For the YECs, the Earth is not 4.5 billion years old but instead 6,000 years old or 10,000–12,000 years old. The discrepancy between 6000 and 10,000–12,000 comes from a disagreement among YEC's regarding possible gaps in the genealogies described in Genesis 5 and 11. If one does not accept these gaps, then 'calculating' the Earth's age is rather easy, as one simply has to add up the genealogies in Genesis 5 and 11 from Adam to Abraham, which comes out to be roughly 2,000 years using the Masoretic Hebrew version. Those 2,000 years are then added to the 4,000 years of time that have passed since the time of Abraham (approximately 2,000 B.C.E.).

In comparison to these three groups, the old-earth creationists appear rather progressive as they accept physics, geology and astronomy and 'only' have issues with biological data pertaining to the origin of species (particularly to the origin of humans). As their name implies, these creationists accept the idea of an ancient Earth. Within this group, though, there are many sub-groups who differ in their means of

---

14    Joshua 10 "The Sun Stands Still" Joshua 10:12… Joshua said to the Lord in the presence of Israel: "Sun, stand still over Gibeon, and you, moon, over the Valley of Aijalon." 13 So the sun stood still, and the moon stopped, till the nation avenged itself on its enemies, as it is written in the Book of Jashar. The sun stopped in the middle of the sky and delayed going down about a full day. 14 There has never been a day like it before or since, a day when the Lord listened to a human being. Surely the Lord was fighting for Israel' (NIV).

incorporating the idea of an ancient Earth and their belief in special creation, i.e., that God directly created the Earth and all life upon it.

One subset of old-earth creationists known as 'gap creationists', for example, believe that the six days of creation are in fact 24-hour days, but they have found an easy solution to incorporate the idea of 'days' of creation with an ancient Earth simply by deciding that there must have been a large temporal 'gap' between Genesis 1:1 and Genesis 1:2. In other words, God created the heavens and the Earth and then took a very long break. After this temporal gap, the 24-hour periods continue in their normal literalist fashion, i.e., there are no more temporal gaps between any of the following verses in Genesis 1:2–1:31 (Scott 2009).

The so-called 'day-age creationists' take a different approach to their solution in that they do not try to incorporate the six days of creation into the modern concept of a 24-hour period but instead simply state that the 'days' of creation were very long, even thousands to millions of years long. This view allows this group of creationists to enjoy their perceived parallels between organic evolution and Genesis, although they ignore the anomalies such as birds occurring before land animals (ibid.).

'Progressive creationism' is the term given to the majority view of today's old-earth creationists. They have basically accepted that simple, single-celled organisms appeared before multicellular organisms, followed by more complex forms of life, much like is seen in biological evolution, yet the progressive creationists differ greatly from theistic evolutionists in that they do not believe that the series of appearance is due to natural or biological evolution but instead that God actively and directly created special kinds as they currently exist.

Where intelligent design should be placed on this spectrum is debatable. I have placed it as the most allegorical (Figure 9), while Scott has placed it outside of the continuum (Figure 8). In essence, both portrayals are correct, and both reflect Phillip E. Johnson's intention in the establishment of the intelligent design movement. As discussed in the chapter on legal cases, intelligent design originated in response to legal losses. Johnson, as a Berkeley law professor, believed that Darwinism could be beaten through simple rhetoric. Johnson recognized that many creationists were divided over their varying interpretations of Genesis, which led to arguments over the age of the Earth, the role that God played in the origin of species, etc. Johnson's idea was to offer a 'big tent' where all creationists could gather around a common belief, namely the existence of a personal God and a dislike for secularism and evolution. So instead of focusing on the Old Testament and the detailed accounts found in Genesis, Johnson moved the focus to the Gospel of John, "In the beginning was the Word (John 1:1)". So, as we can see, intelligent design acts as a 'big tent' as portrayed by Scott but it is also the least literalist strain of creationism as seen in Figure 9. Johnson's strategy is ingenious as his 'tent' does in fact attract many individuals who would not have otherwise ever considered themselves to be a creationist. As various polls in the United States have shown, anywhere between 70% to 90% of Americans believe in God, and potentially there is enough flexibility and space within this 'intelligent design tent' to accommodate everyone from fundamental flat-earth creationists to mainline Protestants.

While Scott and I both included 'theistic evolution' within our spectrums, it should not really be considered a sub-type of creationism. Theistic evolution adherents are generally capable of harmonizing their belief in God with evolutionary theory and unlike the creationists, they do not believe that there is an inherent conflict between religion and science (Numbers 2006). This difference is highlighted by Miller et al. in their benchmark publication on public acceptance of evolution, "The biblical literalist focus of fundamentalism in the United States sees Genesis as a true and accurate account of the creation of human life that supersedes any scientific finding or interpretation. In contrast, mainstream Protestant faiths in Europe (and their U.S. counterparts) have viewed Genesis as metaphorical and—like the Catholic Church—have not seen a major contradiction between their faith and the work of Darwin and other scientists (Miller et al. 2006, p. 765)." Theistic evolution is in fact the official position of the Catholic Church as Pope John Paul II stated in 1996 that God created the world, evolution happened and humans may have descended from more primitive forms, but God was responsible for the creation of the human soul (Scott 2009). In other words, theistic evolution is not concerned with how the physical body came into being but is instead focused on the origin of the human soul. This separation between the origin of the soul and the origin of the physical body makes it much easier for theistic evolutionists to accept the idea of descent with modification through natural processes and thus they are less encumbered in the classroom as they do not feel compelled to reject evolution. For this reason, theistic evolution does not affect an individual's potential scientific literacy.

The very broad spectrum of creationist thought illustrates the vast number of varying preconceptions that students may possess before they even receive their first formal lesson on evolution. Moreover, there is another overarching misconception that precludes and supports the tenacity of these ideas and that is the misconception that religion and science are incompatible and that there is something 'inherently' dangerous about the theory of evolution. We will look at this particular misconception and how it may be remedied in a later chapter, Chapter 8: Understanding Common Descent. Now we will look at how such misconceptions can be categorized in order to understand how responsive or resistant they are to educational intervention.

## Misconception Classification

Before we begin, it is important to point out that experts do not always agree with how creationist thought should be classified. While many authors refer to creationist ideas as misconceptions, Michael Reiss, Professor of Science Education at University College London and former director of science education at the Royal Society, argues that creationism should be considered a worldview and not a misconception. As I have understood it, Reiss' reasoning is that by referring to creationism as a worldview, it highlights the tenacity of the belief and the fact that it cannot be easily rectified through education. Unfortunately for Reiss, his speech on this topic at the 2008 British Association Festival of Science caused some attendees to believe that Reiss was calling for creationism to be taught or at least tolerated as a valid alternative in the science classroom. This interpretation of Reiss' comments caused a group of Nobel laureates to call for Reiss' resignation. While I understand Reiss' reasoning, I will continue to use the word misconception to highlight the fact that when creationist beliefs regarding the shape, age and position of the Earth are based on biblical interpretations instead of scientific fact, then they are in fact incorrect conceptions of the natural world. The world is not 6,000 years old, it is not flat, and it is not the center of the solar system. Likewise, all of life was not created within one week and species were not all created separately as special kinds. To claim otherwise is simply incorrect.

Where I do agree with Reiss is with regard to individuals who hold theistic interpretations of evolution—in other words, those individuals who understand and accept biological evolution and the natural mechanisms that lead to the origin of new species, yet still maintain their own interpretation of God's personal relationship to man. In other words, they rely on science to understand the natural world and how it came into being, while deferring to their faith and religious teachings to understand the spiritual world. In this particular case, the term worldview is a much more appropriate than misconception as a description for this type of thinking because it is not the role of science to interpret the spiritual realm of life or to negotiate the relationship between a person and their chosen god. As Stephen Jay Gould explained more than twenty years ago, "The lack of conflict between science and religion arises from a lack of overlap between their respective domains of professional expertise— science in the empirical constitution of the universe, and religion in the search for proper ethical values and the spiritual meaning of our lives (1997, p. 18)." For that reason, I would like to express that when I use the term misconception, I am referring to those beliefs about the natural world that are categorically false and in contradiction with scientific data.

Now that the use of the term misconception has been clarified, I would like to propose a categorization scheme to highlight the variety of misconceptions that individuals may have about evolution and then discuss how these misconceptions arise and how they respond to educational intervention.

## Categorization of Misconceptions According to Form and Persistence

0. *Neutral (natural) misconceptions*—ideas about the natural world or science that a person has either constructed themselves or attained through the false assimilation of information. Example: Assuming according to one's own cognitive reasoning that the word *theory* is used in science in the same way that it is used in everyday life. This false information is not necessarily taught directly to person, but they may develop this type of misconception simply by assuming a scientific theory is equivalent to an educated guess. This type of misconception could subsequently cause a person to believe that the theory of evolution is simply a 'good guess' but not necessarily true. Another example of a neutral or natural misconception is the false assimilation of information such as a young child confusing metamorphosis with evolution (Figure 7). These types of misconceptions can be understood as a type of 'misunderstanding' due to a general lack of correct information or a lack of intellectual maturity. In this case, the individual is uninformed and thus incapable of comprehending the true complexity of the concept of evolution. These types of neutral (natural) misconceptions are often easily remedied once the person has

obtained the necessary intellectual maturity and is provided with a clear explanation of evolution.

1. *Seeded misconceptions*—ideas about the natural world that are actively taught to an individual by a non-scientific source. These ideas are not necessarily intended to replace scientific explanations but are instead often taught as supplementary information during religious education. Example: A child learns about Adam and Eve or Noah's Ark at Sunday school or from religious free-choice learning materials (e.g., children's Bible). While these ideas are in fact 'taught' to the child or student they are can be taught in a way that is metaphorical or allegorical and thus do not hinder a person's ability to learn about evolution and accept it as a valid scientific theory. According to this classification, a seeded misconception signifies that while the 'seed' of a misconception is present or planted, it has not taken root, meaning that this individual is still quite able to recognize non-scientific teachings about the natural world as allegorical stories and understands that while they may be important aspects of religious learning, they are not correct interpretations of the functioning of the natural world. Often times, this allegorical understanding of biblical creation can be harmonized with scientific data learned later in school as in the case of supporters of theistic evolution. Example: A student who has learned about the Garden of Eden is later able to incorporate these lessons into studies of evolution by seeing Adam and Eve as allegorical *representatives* of the first hominids or Homo sapiens with self-consciousness.

2. *Propagated misconceptions*—ideas about the natural world that are actively taught by non-scientific sources with the clear intent to convince individuals that these are accurate explanations of the natural world that are to supersede scientific explanations. The intent here is to persuade the child or student that the religious explanation of the world should take precedence over the scientific explanations, leading to the rejection of all scientific data which contradict these religious teachings. This is characteristic of groups that emphasize a strict literal interpretation of Genesis, especially those groups who also attempt to use their own version of 'scientific' reasoning to support their fundamentalist ideas, such as flat-earth, geocentric and young-earth creationists. The result of this type of propagated misconceptions is a generalized confusion with regard to the nature of science and the relationship between religion and science. Students with propagated misconceptions perceive an insurmountable conflict between their own faith and what it taught to them in a science classroom—leading to perfunctory rejection of evolution and any other scientific data that contradict the literal interpretation of biblical accounts. Example: publications such as *Galileo Was Wrong: The Church was Right* by Robert Sungenis and Robert Bennett, actively and intentionally propagate the idea that Earth is the center of the universe. In his books and corresponding website, Sungenis uses 'science' to explain the 'validity' of geocentric claims. While the fallacy of their claims is quickly identified by a scientifically literate individual, their argumentations and (mis)use of science and scientific terminology can be very misleading to individuals who do not understand enough about physics or astronomy to discern the falsehood of these geocentric claims. This is also typical of young-earth creationist groups, such as Answers in Genesis (AiG), which produce materials and museum exhibits that actively teach

children about the 'scientific evidence' of a young Earth and even that humans and dinosaurs lived side-by-side (Figure 10). These misconceptions are clearly more obstinate and are often resistant or immune to educational interventions.

3. *Internalized (self-perpetuating) misconceptions*—internalized misconceptions are often the result of either seeded or propagated misconceptions that have taken root within an individual and have become a central component of this individual's self-theory[15] and world-theory. These misconceptions are the most resistant and unresponsive to educational intervention because they have been internalized and are seen as absolute truth. These misconceptions develop into a central component of a person's fundamentalist 'meaning system' and thus play a major psychological role in the manner in which they interpret and respond to particular experiences and life situations (more information on meaning systems will be provided in the next chapter). Due to the link between these types of misconceptions and the individual's sense of self, these individuals begin to actively perpetuate these misconceptions by (1) vigorously rejecting any information that contradicts their view on the 'truth' while also (2) actively pursuing information that supports the maintenance of their distorted view.

These misconceptions cannot be overcome through a simple influx of knowledge in the form of science education as it is impossible for these individuals to harmonize their literalist views with scientific data, e.g., the Earth cannot simultaneously be 6000 years old and billions of years old. Such an individual is thereby convinced

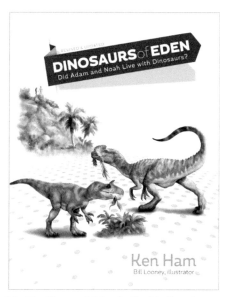

**Figure 10:**  Children's book by Ken Ham explaining the 'truth' of how humans and dinosaurs co-existed.

---

[15]  Self-theory comprises four factors of self: self-image (how a person sees themselves), ideal self (how they believe that they should be), looking-glass self (how they believe that they are perceived by others), and real self (self-image that has been adapted according to feedback from social connections).

that they have only two choices: (a) abandon their beliefs or (b) reject scientific explanations of the world. Due to the deep-seeded psychological impacts of such belief systems (as we will discuss later in more detail) these individuals will almost always choose choice b. This is understandable as there is an absolute lack of intrinsic motivation for a person to abandon their own sense of truth (that they have held for many years or a lifetime) in order to accept scientific data that is often presented to them by a person outside of their circle of trust. Example: A child receives lessons on the 'scientific' reality of a young Earth from a trusted source, such as a parent or grandparent. He learns about special creation and how herbivore dinosaurs lived together with humans (Figure 10). These lessons are internalized and become the foundation of his understanding of the natural world. They also become foundational in his understanding of a personal God and his relationship to his God and act as the basis for many valued social connections. Every lesson that he receives about an old Earth or about dinosaur fossils that are millions of years old is perceived as a direct assault on his faith, his understanding of the world and his personal relationship to God. He thus actively avoids situations and individuals who appear to him as a threat and instead seeks the company of like-minded individuals. Through this process of distancing himself from 'non-believers' he develops ever increasing tendencies towards fundamentalist thinking and behavior.

## *Educational Intervention*

Internalized or self-perpetuated misconceptions clearly pose the greatest hinderance to scientific literacy since the individuals who harbor these misconceptions will actively reject all information that contradicts their idea of 'truth'. This is in stark contrast to neutral or natural misconceptions that are easily rectified through education (Figure 11).

The internalization of these misconceptions that are based on a literal interpretation of the Bible is in essence the defining move from religious belief to

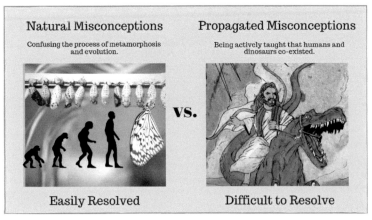

**Figure 11:** Examples of children's misconceptions about evolution that arise naturally versus those which result from the active teaching of scientifically false principles (Photos from flickr and pixabay).

fundamentalism. If we look again at the definition of fundamentalism and compare it to the Oxford dictionary's definition of faith and religion, we see that the clear distinction is the reliance and adherence to a literal interpretation of a religious text.

**Faith:** *(noun) 1. Complete trust or confidence in someone or something. 2. Strong belief in the doctrines of a religion, based on spiritual conviction rather than proof.*

**Religion:** *(noun) 1. The belief in and worship of a superhuman controlling power, especially a personal God or gods. 2. A particular system of faith and worship. 3. A pursuit or interest followed with great devotion.*

**Fundamentalism:** *(noun) 1. A form of a religion, especially Islam or Protestant Christianity, that upholds belief in the strict, literal interpretation of scripture: Modern Christian fundamentalism arose from American millenarian sects of the 19th century, and has become associated with reaction against social and political liberalism and rejection of the theory of evolution. [emphasis added]*

Moreover, internalized misconceptions about the natural world are also often accompanied by misconceptions about the nature of religion. In other words, creationists who deny the old age of the Earth would also most likely assert that religion and science are inherently in conflict. This combined misunderstanding of both the nature of faith and the nature of science further complicates the situation and often leads to perfunctory rejection of evolution. When being taught about evolution, students with internalized creationist misconceptions perceive themselves as a person of faith who must uphold their religious beliefs against scientific attacks.

In truth, though, these students do not need to choose between religion and science but instead they are faced with the decision between fundamentalism and religion. In other words, they must choose whether their faith and their relationship to God is dependent upon a literal interpretation of the Bible or if they can base their faith upon an allegorical or metaphorical interpretation of the Bible (Figure 13 and Figure 12). Thus, part of the solution to this situation will also require that individuals are capable of better recognizing the difference between religion and fundamentalism and realizing their own position on the spectrum between faith and fundamentalism. It is also necessary for teachers to keep this key distinction in mind as not all religious students are fundamentalists.

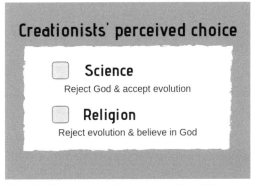

**Figure 12:** Creationists' perceived necessity to choose between their faith and evolution.

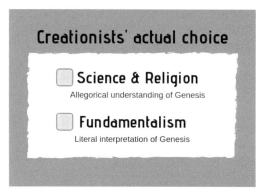

**Figure 13:** Creationists actually only need to decide between of fundamentalist view of the Bible, i.e., literal, or allegorical as proposed by mainline Christian leaders and theologians.

From an educational standpoint, it is important to keep in mind that while we are interested in increasing scientific literacy and curiosity, we cannot overstep the boundary between education and indoctrination. Just as we hope that students will recognize that religion and science are two separate magistrates, we too must acknowledge the students' right to understand their world from both of these realms. This is important from a legal perspective as well as from psychological and educational perspectives.

From a legal stand point the U.S. Constitution clearly preserves an individual's right to practice religion—meaning that a student cannot and should not be forced to forfeit their religious beliefs or faith in God. When it comes to teaching evolution, the law is on the side of the secular science teacher, and a teacher has the right to prevent the discussion of creationism in the classroom due to the separation of church and state. If students challenge the teacher, the teacher also has the right to point out the difference between faith-based ideas such as intelligent design or creationism and evidence-based scientific theories such as evolution. At the same time, it is important that teachers act conscientiously when approaching the subject of evolution to avoid unnecessary litigations and undue stress, i.e., that teachers refrain from insulting the student or suggesting that their religious ideas are improper. The only legal case that involved a teacher being sued for 'teaching' evolution was a history teacher, not a science teacher, and he was sued for calling creationism "superstitious nonsense". As described in the past chapter, the court found in favor of the prosecution stating that the teacher's comment regarding creationism lacked a secular purpose and thus "constitutes improper disapproval of religion in violation of the Establishment Clause".

In truth, I believe that cases like these and possibly all litigation can be avoided simply by (1) having a trusting relationship to the students and (2) acting respectfully towards them, regardless of their views and beliefs. All of the classroom tools that I offer in part two of this book are all ethical and legal means of addressing fundamentalism in order to support a student's acquisition of scientific understanding without going to court. This respectful, relationship-oriented approach is not only a useful means of avoiding litigation but also makes sense from an educational standpoint. The last thing we want to do is to confront students with an either-or choice. In other words, we must guard against the situation where students feel as if they must choose between two equally unattractive options (a) *either* you accept evolution and abandon your faith *or* (b) you maintain your religious beliefs and remain scientifically illiterate. Remember, this is the idea that creationist students often bring with them to the classroom and we do not want to reinforce it. Moreover, as we will discuss in the next chapter, religion and even fundamentalism play central roles in the development of 'meaning systems' and 'terror management strategies' and it is therefore also important that we do not try to persuade a person to give up their religious beliefs as we cannot be sure how their faith is involved in their ability to manage difficult life situations.

Thus, for legal, psychological and educational purposes, students must have the option and opportunity to become scientifically literate while still maintaining their chosen faith (Figure 13). For this to occur, though, they will need to learn to recognize and relinquish literalist thinking, i.e., fundamentalist tendencies. So how do we accomplish this very delicate balancing act?

As Keith S. Taber, professor at Cambridge university has clearly stated, "Science teaching is not about persuading students to believe things (2017, p. 53)." Instead science education should be focused on assisting students develop an ability to 'think scientifically' about the natural world. According to Taber, it actually boils down to teaching students what and how to *doubt*. In other words, science education should help students think critically while fostering curiosity for the natural world and an understanding of how hypotheses are created and how they are tested. Good science education should also provide students with the opportunity to develop their own ideas as well as the time and space to test these ideas using scientific methods. The good news is that this type of education can begin even at a very young age.

Simple projects are often enough to get the ball rolling. For example, a group in Germany set out to examine how even young children could learn about the nature of science. Within the framework of their research, a group of children was asked what happens to the apple seed when the apple itself rots. There were many different answers: Some of the children believed that the seeds rotted as well, some believed that a tree would grow from the seed within the rotting fruit and some believed that a new apple would emerge from the seed. The task of the educators was then to guide the children in finding ways that they could empirically test their ideas in order to find the answer (Masuch et al. 2007).

These type of simple learning experiences offer even young children the tools they need to develop and test hypotheses instead of making and maintaining (false) assumptions about the natural world. As children progress through the school system, this type of scientific thinking can be expanded upon through lessons on the scientific method and the execution of more complex experiments. Moreover, it is vital that students are taught and have the opportunity to practice systematic literature reviews in order to learn the difference between obtaining reliable scientific data from peer-reviewed, scientific sources versus unsubstantiated information from blogs, social media or other online sources.

In this way students develop the knowledge necessary to recognize the difference between assumptions/beliefs and facts/data. Moreover, they have the means to test or check the validity of their beliefs/assumptions. In this way students grow up to be scientifically literate citizens who are capable of recognizing when they have a false perception about an aspect of the natural world and offers them a means of remedying this misunderstanding themselves through experiment, observation or a systematic review of the literature. While we cannot expect students or the general public to automatically rectify all of their misconceptions themselves, a few key experiences do provide the them with a taste for scientific endeavor and encourages scientific thought patterns.

Of course, the situation becomes exponentially more complex when emotions are involved. When a child realizes that a tree grows from a seed and that apples do not directly emerge from seeds, we can expect that the child is able to accept that their original idea was wrong and rectify it without any dramatically negative emotions. This is even true when the information that is presented is counterintuitive such as the fact that tomatoes are technically a fruit. While this may cause the unpleasant feeling of 'being wrong' while the other students were 'right', this feeling is probably best equated to the negativity associated with a losing a game but not of emotional trauma.

Trauma may sound like a harsh word, but the unraveling of misconceptions, such as the age or location of the Earth, may in fact be best described as traumatic rather than simply uncomfortable—especially when we are addressing beliefs that are deeply entangled with a person's sense of self and their understanding of their role in the world and their relationship to their God. As Miller et al. wrote, "Evolution is nonetheless problematic to some of these nonliteralist Christians, because it implies a more distant or less personal God (2006, p. 765)." So, we can image that students who come to the class with much deeper-seeded, propagated or internalized misconceptions, will feel much more uneasy about evolution and most likely perceive it as a threat. For example, if a person has been convinced for their entire life that they inhabit a planet that is less than 10,000 years old and that all of life on this planet is the product of special creation, then finding out that the world is actually billions of years old and that all of life evolved from primitive single-celled organisms is clearly going to illicit a negative emotional response. This is very understandable as the information presented to them in direct conflict with everything they have known since they were a child. It is not simply another 45-minute lesson in the classroom but a test of their faith as it directly contradicts the literal basis of their belief system.

While many experts may continue to debate Michael Reiss' proposition that creationist beliefs should be categorized as worldviews instead of misconceptions, the implications for science education remain the same whether we call these beliefs worldviews or misconceptions. If a creationist student is presented with a lesson on human evolution that clashes with their literal interpretation of Genesis and thus the foundation of their belief system, then it means that they can either (a) find a means of integrating the new information by restructuring their previous belief system or (b) reject the science presented to them. Here it is clear that choice (a) is the preferred choice by science educators, but also the more difficult option for the creationist from both an intellectual and emotional point of view.

The dismantling of these misconceptions is a vital step towards developing the ability to comprehend scientific data. Yet, the unraveling of creationist misconceptions cannot be seen as being equivalent to undoing any other educational misconception. For conceptual change to occur, a person must first see that their initial view is incompatible with what they are being taught and then *choose* to accept the new information by rejecting or reworking their previous beliefs. This process is severely complicated by the involvement of emotions and personal identification, as is the case with creationist thinking. The table below offers a few examples of preconceptions that are challenged by new data. While some of these points involve a great deal of emotion, others result in almost no emotional response or possibly only confusion (Table 2).

As can be seen from these simplified examples, a creationist being confronted with evolution cannot be equated with rectifying misconceptions about geography or botany. Instead, we must understand that the emotions involved in confronting this misconception is much more similar to that of a husband finding out his 'perfect' wife is cheating on him. The reason for this is simple. Our 'belief' in the stability and permanence of human relationships is also deeply integrated into our sense of self and our place in the world. Likewise, creationist beliefs are also a major component of creationists' identity and form the basis of their personal relationship to their God. It is therefore understandable that when this system of belief is challenged that the individual reacts to the information in a manner that is similar to that of an individual going through the grief process.

According to the Kübler-Ross model there are five 'typical' stages of grief that appear to be natural emotional responses that occur when humans are confronted by sudden, devasting news, such as the diagnosis of a terminal illness (Kübler-Ross 1969). These stages are denial, anger, bargaining, depression and finally acceptance (Figure 14).

Looking at these five stages, it is easy to recognize the parallels to stages of the creationist movement discussed in the first chapter. First, creationists reject the theory of evolution (denial), then angrily try to keep it out of the classroom (anger) and finally try to bargain by proposing that we 'teach both sides' (bargaining) (Table 1). This process appears to have parallels not only for the creationist movement but also for the creationist on a personal level as they move through various cognitive

**Table 2:** Potential emotional responses when preconceptions are contradicted by new data.

| Preconception | New Data | Emotional Response | Possible Feelings |
|---|---|---|---|
| Believing tomatoes are vegetables | Finding out tomatoes are technically a fruit | Little to none | Confusion |
| Believing that the tooth fairy brings money for old teeth | Finding a collection of old teeth in your parents' closet | Moderate. Age dependent | Shock or disappointment |
| Believing Dallas is the capital of Texas. | Finding out Austin is the capital of Texas | Little to none | Shame or confusion |
| The belief that a relationship will last forever | The realization that the relationship is over | High | Anger to rage, sadness to depression, denial, or mix. |
| The belief that all life on earth is a product of special creation | The fact that all life on Earth developed from primordial soup | High | Shock or sadness or anger or denial or mixed |

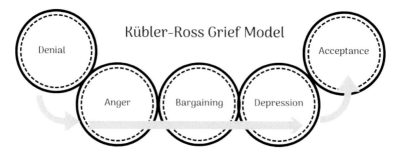

**Figure 14:** Kübel-Ross Grief Model.

**Figure 15:** Potential parallels between Kübler-Ross Grief Model and the creationist movement (both societal and personal).

and emotional stages on their way to accepting scientific facts. While our ultimate goal is to have creationists accept the theory of evolution as a valid scientific fact, it means that they will likely transgress through a state of depression (or deep sadness) on their road to acceptance (Figure 15).

My goal in pointing out this apparent parallelism between the grief process and the path a creationist takes in coming to terms with evolution, is to highlight the emotional impact learning about evolution can have on creationists. So, while experts may still debate over the classification of creationism as a misconception or a worldview, I would like to propose that it is best to think of creationism as a misconception from an intellectual perspective and a worldview from an emotional perspective. In terms of education it would be important to correct the misconceptions of the natural world, while also respecting the emotional impact that this teaching will have on the student as it influences their understanding of the world. Understanding this duality is essential, so that we do not underestimate the emotional repercussions and ramifications that the correction this misconception will have on the student.

# Summary

Many misconceptions about the natural world are harmless as they are simply the product immature knowledge about a certain subject, yet creationist beliefs represent a major hinderance to scientific literacy because they are not based on simple misunderstandings or a lack of information but are instead deeply entrenched in the person's sense of self and their understanding of the world around them. While there is a wide spectrum of creationist thought and varying degrees of literalism among creationist groups, all of these beliefs are intrinsically connected to the creationists' personal and cultural identity. What this means is that the discussion of human origins is not a purely intellectual issue for creationists but a far more emotional subject. It is thus clear that unraveling these misconceptions cannot be equated with the simple correction of false knowledge, which means that there are no easy or superficial solutions.

In order to find a solution to this conflict, we will thus continue delve even deeper, beyond the effects of our own personal, family and national heritages, in order to discern how we are also influenced by our evolutionary history. In the upcoming chapter we will examine this conflict situation from an evolutionary stand point and explore possible evolutionary causes for creationist belief systems both at the individual and societal level.

# Chapter 4

# Our Inner Neanderthal

The title of this book *Neanderthals in the Classroom* was chosen to highlight two important aspects of the complexity of the creationism | evolution conflict. The first point is that the largest problem that opponents of evolution have with the theory is the involvement of humans in the concept of common descent. Had Charles Darwin written a book entitled *On the Origin of (non-human) Species*, his ideas would have most likely not received the same attention or caused the same degree of controversy as it has. Even prior to publishing his ideas on organic evolution, Darwin understood how controversial the topic of *human* origins would be and thus intentionally skirted the topic in his original 1859 publication, including only one brief statement saying, "Light will be thrown on the origin of man and his history (p. 488)." Despite the subtleness used to express his ideas regarding human origins, the implications were clear. The idea of an organic human origins was seen as a threat to core religious ideas and many interpreted Darwin's theory as a degradation of a personal relationship to God. In effect, Darwin's theory dethroned human beings as the crown of creation.

Once his initial theory of evolution had gained general acceptance in academic circles, Darwin decided to address human evolution explicitly and in great detail in his 1871 publication, *The Descent of Man*. The idea that modern humans are simply the product of mutated primitive beings upset many and was seen as a serious threat to the core of humanity (Kindt and Latty 2018). Regardless of how despicable the idea of organic human evolution may seem to some—the fact remains that it is simply true.

Thanks to modern science and genetic research, it is no longer possible to deny our primitive past. Studies from 2010 first showed that people of Eurasian origin have inherited anywhere from 1.5 to 4 percent of their DNA from Neanderthals and more recent studies from 2016 have shown that these archaic hominid alleles are associated with certain neurological, psychiatric, immunological, and dermatological phenotypes (Simonti et al. 2016). In other words, our inner Neanderthal still has observable effects on our brain chemistry and physical health.

This points to the second idea behind the title of this book which is to highlight the effect of our own evolutionary past on our current lives and the manner in which we engage with one another. In other words, the rejection of evolution and the conflict over human origins is ironically affected by our own evolutionary instincts that occur well below our level of consciousness. In this chapter, will examine evolutionary causes and implications for (1) why people intentionally uphold false beliefs about the natural world, (2) how threatening or questioning these belief systems can cause a fight-or-flight response, and (3) how our tendency to 'other' exacerbates the conflict and prevents learning from occurring. We will see how point #1 is linked to inherent pleasure seeking and pain aversion tendencies, point #2 relates to threat recognition behavior, and point #3 relates to tribal behavior.

## Evolutionary Causes of Creationist Clinging

*"Nothing in biology makes sense except in the light of evolution."*

Theodosius Dobzhansky (1973)

When I first began to study the phenomenon of creationism, I was often asked by my friends and colleagues, "But how is it possible for anyone to truly believe in a young Earth or a flat Earth when we have access to so much data that contradict this belief?" The answer to the question is easy: Because they want to. While my friends and colleagues most likely expected me to say that creationists are just ignorant or unintelligent, this is simply not the case. Creationists are not convinced of a young Earth because they are *unaware* of the scientific data available on the age of the Earth nor do they reject evolution because they have never received information on the validity of the theory of evolution. Creationists hold these beliefs despite access to this information because they actively and purposefully seek out other sources of information that support their literalist views while attempting to find loopholes in the scientific data that substantiate evolution.

This in itself is not surprising as it is well known that individuals prefer information that affirms their own beliefs or confirms what they would like to believe is true. So, the more interesting question then becomes—*Why?* Why would creationists want to hold on to a literal belief in the Bible? In order to answer the question, I decided to take another look at creationists, but this time from a different perspective—from the standpoint of a biologist. Instead of only looking at *creationism* solely as a cultural or societal phenomenon, I began to look at the *creationist* from an evolutionary standpoint. As a biologist, I had learned that for every 'bizarre' trait or behavior, there was often some logical evolutionary cause for the appearance of such a trait or behavior. I began to ponder whether or not there was an evolutionary advantage or cause for creationist thought and behavior. Just as biologists were able to explain the peacock's strange tail through sexual selection, might there also be any evolutionary explanation for this 'strange' data-denying behavior?

This analysis also allows a look at the creationist as a person independent of the creationist movement. While we have established how creationism has developed in the United States, I wanted to understand the creationist's motivation to adhere to creationist beliefs. This analysis should not be misunderstood as an attempt to rationalize or legitimize this type of fundamentalist thinking, nor is it meant to patronize the creationist. Instead I was inspired to pursue this analysis because I became convinced that by identifying the biological (evolutionary) explanation for this behavior, it may be possible to offer a better understanding of creationist clinging and thus find more effective ways of engaging with creationist students. While many continue to be convinced that this clinging to 'nonsensical' ideas is the product of ignorance, it is far more complex than that. It is important to move away from this idea of 'ignorance' in order to recognize that creationist thought is a choice and ideally, to understand the reasons behind that choice. By recognizing this difference between ignorance and choice, we can see why the key to developing effective science communication cannot simply be found in the presentation of more facts. So, in our attempt to develop more effective means of engaging with creationists, we will first look at the possible evolutionary reasons behind their choice to cling to fundamentalist ideas.

Evolutionary psychology is a theoretical approach to psychology that examines psychological mechanisms from an evolutionary perspective with the central aim of identifying which psychological traits can be considered adaptations, i.e., functional products of natural or sexual selection in human evolution (Racevska 2018). According to experts, such as David Buss from the University of Texas, evolutionary psychology is not a distinct branch of psychology, but rather a "theoretical lens that is currently informing all branches of psychology (1998)." While adaptationist thinking is common in evolutionary biology with regards to physiological mechanisms, such as the immune system or sweat glands, evolutionary psychologists apply this same type of adaptationist thinking to psychology, in an attempt to understand the adaptive psychological benefits of particular human behaviors such as religious belief. Many evolutionary psychologists, in fact, argue that a great deal of modern human behavior is the product of adaptations that developed in response to recurring problems in our ancestral environments (Buss 1995).

After conducting a survey of the literature on the psychological role of religion and even fundamentalism in a person's life, I believe that there is in fact reason to believe that there are evolutionary causes that explain creationist clinging and that they are linked to latent survival structures in our brains. Here I outline what I see as some of the most basic underlying causes for this behavior: (1) avoiding pain while seeking pleasure, and (2) human need for connection.

## Fundamentalist Beliefs as a Means to Avoid Pain and Seek Pleasure

Even the simplest of organisms move away from painful experiences and move towards pleasurable experiences—a general behavior that normally assists in the increased likelihood of survival. This pain avoidance or aversion behavior is seen at the level of paramecium responses to toxic substances, as well as the neurological wiring responsible for unconscious reflexes that respond to pain, e.g., the reflexive removal of a hand from a hot stove. Seeking pleasure is also a central survival mechanism that has evolved to encourage us to find food, procreate, take refuge from the cold, etc.

The brain's primary role is to maintain a certain balance in the body's internal environment and the brain achieves this by giving instructions to the body on how to respond to the environment to correct any imbalances as they arise. In 1970, Jeffrey

Alan Gray proposed one of the most widely accepted theories in terms of biological models in psychology known as the biopsychological theory of personality, which describes the existence of two systems that drive behavioral activity, namely the behavioral inhibition system (BIS) and the behavioral activation system (BAS). According to this theory, BAS involves the pursuit (approach) of positive stimuli and is linked to positive emotional states, while the BIS involves the avoidance of negative stimuli and is linked to negative emotional states (Gable et al. 2000).

Despite the hundreds of millions of years that lie between us and our single-celled organism origins and even the 50,000 years since the emergence of behavioral modernity, there is still a strong drive within us to seek pleasure and avoid pain. Yet, while these neurological systems supported primitive humans in their ability to survive, they can also have dramatic, negative consequences in our modern society, e.g., our attraction to sugar and fats, which helped our Neanderthal counterparts survive, has led to an obesity epidemic in our modern society.

This approach and aversion behavior holds true not only in terms of aversion to physical pain and the seeking of physical pleasure but is also true of psychological pain and pleasure. Here we will attempt to answer the question of whether clinging to creationist beliefs can be seen as a form of this pain aversion | pleasure seeking behavioral pattern. In other words, can creationist clinging to literalist beliefs be explained by the fact that these belief systems bring them pleasure in the form of positive (pleasurable) emotional states, while simultaneously ameliorating negative (painful) emotional states?

According to many studies, religious adherence and belief is in fact linked to increased physical health and social support as well as a greater sense of well-being (Luhrmann 2013). For this part of the analysis, we will focus on the psychological benefits, as it has been shown that the psychological benefits persist regardless of whether these individuals are adherents of mainline religious groups or fundamentalist fractions (Hood et al. 2005).

Before we continue, though, I would like to reiterate that the immediate goal of this analysis is to better understand the underling evolutionary drive to self-perpetuate literalist beliefs in a young Earth despite clear scientific data to the contrary. This analysis should in no way be understood as a support of fundamentalism or any attempt to undermine or validate any particular religious belief or creed. The point here is that irrespective of the validity of any religion's supernatural claims, there are statistically proven psychological benefits of religious faith and even fundamentalist worldviews. In the following section, I will limit myself to these studies as anything further would over step the boundaries of a scientist or an educator. In later parts of this book, we will discuss whether or not these same psychological benefits can be achieved without the dogma and literalist clinging that cause cognitive inflexibility. With that being said, we begin by looking at the psychological aspects of religion in the form of 'meaning systems'.

## *Meaning Systems*

When it comes to discerning the link between increased well-being and religious belief (or fundamentalism), many scientists propose that one of religion's greatest lures is its contribution to a sense of coherence in an otherwise chaotic world by providing an idea of moral certainty and stability (Hood et al. 2005). This idea is supported by research conducted by neurologist and psychiatrist Michael Inzlicht and his colleagues who state that one of the main advantages of religion is the creation of a 'meaning system' (2011). Hood et al. define a meaning system as "a group of beliefs or theories about reality that includes both a world theory (beliefs about others and situations) and a self-theory (beliefs about the self), with connecting propositions between the two sets of beliefs that are important in terms of overall functioning (2005)." Other psychological theorists posit that such meaning systems are necessary for an individual to be able to function in the world, especially when they are faced with difficult situations. While these meaning systems are not always religious in nature, religion may be one of the most satisfying and comprehensive meaning systems (Hood et al. 2005).

In fact, religious belief has been shown to offer such substantial well-being and health benefits to its adherents that the positive attributes can actually be measured at the level of the brain (Inzlicht et al. 2011). Inzlicht et al. propose that these neurological effects may be best explained, at least partially, from a 'motivated meaning-making' perspective (2011). In other words, individuals are motivated to seek and create a sense that the world is an orderly place, and this can be accomplished through a sense of coherency between beliefs, goals and an individual's perceptions of their environment. In this way, religious beliefs can buffer feelings of distress associated with disruptions in the system (Inzlicht et al. 2011).

This idea is further supported by the research conducted by Ralph W. Hood, Jr. and colleagues who stated, "Religion is a worldview; that is, it becomes a primary meaning system through which all of life is viewed and understood (p. 13)" and this appears to be even stronger in the case of fundamentalism, where religion becomes an all-encompassing way of life (Hood et al. 2005). Yet, Hood et al. are clear to point out that the difference between a highly religious person and a fundamentalist is not based on the degree to which religion plays a role in their life but the mode in which this meaning is discovered. In other words, we cannot categorize a person as a fundamentalist solely on their degree of religiosity, i.e., even if religion is the central meaning system in a person's life, they are not necessarily a fundamentalist. Instead, the authors rely on the role of literalism and religious texts in the production of meaning systems to define the difference between fundamentalism and religion, as they describe "… we do not identify as fundamentalists all people who take religion as a primary meaning system. What distinguishes fundamentalism from other religious profiles is its particular approach toward understanding religion, which elevates the role of the sacred text to a position of supreme authority and subordinates all other potential sources of knowledge and meaning (p. 13)."

In other words, the fundamentalists' meaning system is based almost exclusively on the content of their sacred texts and they insist that "all of life be understood in relation to the text (p. 5)." In the case of creationists, we see Protestant fundamentalists who believe that the Bible is the direct and literal revelation of God and that it is therefore sufficient as a source of meaning and purpose of life (Hood et al. 2005).

This emphasis on sacred texts is an important point that must be considered when attempting to elicit conceptual change with regard to creationist beliefs.

Because the creationist's meaning system is derived from a literal interpretation of the Bible and evolution clashes directly with their literal interpretation of Genesis, evolution becomes a direct challenge to the creationist's meaning system. Clearly, the creationist will want to uphold their meaning system due to the psychological benefits they obtain from it. This can subsequently result in a general rejection of all data that contradict the literal interpretation of the Bible.

These fundamentalist views can also have much larger effects on society when the importance of upholding these meaning systems becomes so great that individuals attempt to forcibly transform their environment to conform to these meaning systems. In the most extreme case can be seen in fundamentalist terrorism as an "attempt to transform the world into a religiously ideal world (Silberman 2005, p. 529)" but is also seen in less violent actions such as the formation of the Christian Coalition or the Moral Majority in an attempt to affect change within the American political and educational system to reflect a more 'Christian nation'.

It is important here to point out that while faith, spirituality, religion and fundamentalism can all act as the basis of such meaning systems and all can offer a sense of coherence and associated increases in well-being, the reliance upon literal interpretations in the formation of fundamentalist meaning systems causes them to be much more brittle and inflexible. This brittle nature often leads to an 'either or' view of the world and requires these views to be forcibly defended as there is no grey zone (see Figure 12). We will discuss this reactivity to evolution in more detail in the next section but first we will look at the role of religion and fundamentalism in the development of 'terror management strategies'.

## Terror Management

Another proposed psychological benefit of religious belief is that it functions as a strategy to ameliorate our fear of death by offering both symbolic and literal immortality. Sociologist Peter Marris suggests that humans possess a "deep-rooted and insistent need for continuity" which results in "the impulse to defend the predictability of life (1986, p. 2)." It is undeniable that one of the most painful emotional experiences humans endure is when we are confronted with the impermanence of our own lives or the lives of our loved ones. Like all evolved beings, humans are programmed for survival, but unlike other animals, humans have the mental capacity to understand their own mortality, which often causes anxiety, fear and apprehension. One means of ameliorating these painful emotions is through the belief in an afterlife. Through the belief in an afterlife, our 'life' is then no longer limited to our earthly stay but has the potential of being extended indefinitely—so long as certain rules are followed. This type of belief system can indeed offer a great amount of solace, not only in facing our own death, but in accepting the death of loved ones. Studies on the emotional effects of a belief in an afterlife have shown that both religious attendance and the belief in an afterlife are inversely associated with feelings of anxiety and positively associated with feelings of tranquility (Ellison et al. 2009).

The anxiety that humans experience in response to their own mortality and our intrinsic desire to reduce this unpleasant emotional state is outlined in what is known as terror management theory (TMT). Terror management theory which was originally proposed in 2015 by Jeff Greenberg, Sheldon Solomon, and Tom Pyszczynski in their book *The Worm at the Core: On the Role of Death in Life*, but the ideas of TMT were derived from earlier work by anthropologist Ernest Becker, author of the 1973 Pulitzer Prize-winning book *The Denial of Death*. According to Becker, most human action is focused, consciously or subconsciously, on ignoring or avoiding the inevitability of death (1973). According to TMT, humans are programmed for self-preservation, like all other animals, yet they are also consciously aware of the fact that death is unavoidable and to a certain extent unpredictable. The combination of this innate drive to survive, coupled with the cognitive ability to understand one's own mortality can cause large degrees of anxiety, or *terror* (Landau et al. 2007).

In order to function, this terror must be managed and according to TMT humans have developed symbolic systems that imbue human life with enduring meaning beyond the physical body in order to manage anxiety about their own mortality. According to TMT, these symbolic systems can include religious systems that offer literal immortality (e.g., afterlife or reincarnation) or cultural systems that offer symbolic immortality (e.g., the idea that we are part of something larger that will outlive the individual such as a country, lineage, legacy, etc.). Interestingly, while this appears to be a universal psychological phenomenon, it also occurs almost entirely subconsciously. Meaning that we are all engaging in this type of mental activity and yet we are almost entirely unaware of it or the effect that it has on us.

While the creation of these immorality systems appears to be arbitrary, they do make sense from an evolutionary standpoint. As Mark J. Landau et al. propose in their paper *On the Compatibility of Terror Management Theory and Perspectives on Human Evolution*, TMT begins with many assumptions about human evolution, "TMT starts with Darwin's (1859) insight that human beings, like all other living species, are biologically predisposed in many ways toward continued life, but that more so than other species, humans adapt to their environment and prosper largely by virtue of highly developed cognitive abilities, including the capacities for abstract, symbolic, temporally extended, and self-reflective thought. Presumably these capacities conferred a significant advantage for humans in terms of flexible and innovative behaviors suited to their physical and social surroundings. This cognitive sophistication, however, had some problematic consequences (p. 477)." TMT further posits that humans were able to use the same cognitive abilities to deal with this maladaptive existential anxiety by creating cultural institutions and belief systems that ameliorated this anxiety. TMT thus considers all cultural beliefs that ameliorate this death anxiety as evolutionarily adaptive behaviors that arose to support the functioning and survival of the collective.

Unfortunately, as anthropologist Ernest Becker argues, these immortality systems have also acted as fundamental drivers of human conflict due to the variation and arbitrary nature of these belief systems, coupled with an individual's strong desire to uphold the superiority of their own immortality system (1973). This defensive behavior is to be expected as each party involved has an intrinsic desire to prove that their belief system is the true system, as only in this way can the system serve its anxiety ameliorating function. I believe that this may be one of the root causes, albeit subconscious, of the intense emotions involved in the creationism | evolution debate. Each side, whether consciously aware of it or not, is deeply invested in upholding their own literal or symbolic immortality system. While evolution posits a *symbolic* form of immortality in the form of reproduction and the overall survival of a *species*, creationists are deeply invested in their *personal* and *literal* immortality, which is fundamentally connected to their belief in Jesus Christ and guided through the literal interpretation of the Bible.

By understanding that a literal interpretation of Genesis is the foundation of the creationist's meaning system, one can better grasp why students will actively resist learning about evolution as it threatens to bring down their entire house of cards, from a psychological standpoint. While there is not an inherent conflict between religion and science in general, the theory of evolution is a direct contradiction of the literal interpretation of Genesis. Because this Genesis account is so essential to the creationist's meaning system and their terror management system, it is understandable that lessons on evolution have the potential of destabilizing these systems, thereby throwing the creationist back into a chaotic and unpredictable world where they are also forced to come to terms with their own mortality. One of the clearest personal accounts of this decisive link between religious belief and mortality comes from Jerry Coyne who said, "I was in high school when the Sgt. Pepper's album came out. I was lying on my parents' couch listening to this new album and all of a sudden it just popped into my mind that everything I'd been taught about God and religion had no evidence behind it. I started sweating, but not because of the heat. I always thought there must be an afterlife. And the sudden realization that probably wasn't true made me start shaking and sweating. Ever since then I've been an atheist (Worrall 2015)." While Coyne offers this story as an example of his emancipation from religion, we can imagine that for other students who hold their faith and their idea of immortality dear, that this would be a very troubling revelation. Because it is natural for humans to want to avoid such painful experiences, a creationist's initial and natural response will thus be to reject any teaching that calls the literal interpretation of Genesis into question. This is why it could be argued that creationists do not reject evolution due to ignorance but because they are responding to impulses sent by the oldest structures in their brain that are programed to do exactly this: avoid pain and seek pleasure. In the next section we will discuss how our tribal past and our biological need for connection plays a further role into this complex conflict situation.

## *Products of Our Tribal Past*

Now that we have established that religion, religious beliefs and even the clinging to literalist interpretations of the Bible all make sense in the context of evolutionary psychology due to the mediating effects that literal and figurative immortality can have when contemplating our own death and how these belief systems support adherents' ability to deal with uncertainty, we can now briefly discuss one more evolutionary cause of creationist clinging that is grounded in our human need for connection.

Above all, humans are social beings and while our tribal past may seem very distant from our lives in modern, developed countries—humans are still evolutionarily programmed to belong to a group or to be a member of a community. As William A. Haviland et al. stated in their *Anthropology: The Human Challenge*, "Born naked and speechless, we are naturally incapable of surviving alone. As humans we rely on culture, a shared way of living, to meet the physical, social, economic, and ideological challenges of human survival (2010, p. 322)."

Humans are thus inherently dependent upon one another. Even in today's society we still depend upon one another for our source of food, clothing, housing and daily comforts, even if this interdependence is easy to overlook in our day-to-day lives, where products are ordered from online shops and appear in our mailboxes. Despite our lack of awareness of this interdependence, ancient neurological structures are still deeply rooted within our modern brains that recognize the benefit of a group and have also evolved to *ensure* that we function within a group.

Thus, the necessity of community cannot be overemphasized, especially from a psychological perspective. According to neuropsychologist Rick Hanson, humans have three basic core needs: the need for safety, for satisfaction, and for connection (2013). Most important for our current discussion here is the fact that this core need for *connection* can often be fulfilled through membership to a religious group. More abstractly, it can also be seen in the development of a 'personal' relationship to God or Jesus, where the individual feels heard, cared for and loved. As the Irish poet John O'Donohue writes, "Prayer is the voice of longing. It reaches outwards and inwards to unearth our ancient belonging (2009)."

For this analysis, though, we will focus on the connection to other humans in the form of a religious (fundamentalist) community. Michael Reiss describes this as the social and institutional dimension of religion. The social dimension of religion relates to its collective manifestation, where members identify as belonging to a particular community of believers or adherents, such as the Sangha in Buddhism, the Umma in Islam and the Church in Christianity (2008). This development of community may be one of the most compelling reasons why religion has persisted throughout human history, as professor of human evolution Lyle B. Steadman and anthropologist Craig T. Palmer posit that one of the most important and significant characteristics of religious behavior is its ability to create "enduring family-like cooperation between non-family members (2008)."

This sense of community and the benefits of belonging to a social network have even been credited with many of the physical benefits of religious adherence such as increased longevity. These benefits have been illustrated through a number of studies such as a 1999 study that showed that weekly religious attendance resulted in seven additional years of life expectancy (Hummer et al. 1999); a 1997 study found that weekly church attendees were 25% less likely to die during the observation period (Strawbridge et al. 1997); and another 1999 study involving individuals over 64 found a 46% reduction in death hazard for frequent church attendees (Koenig et al. 1999). Each of these studies attributed these increases in longevity to the benefits of social ties, expanded social involvement and an increase in social support.

While this development of community is truly a positive aspect of religion, it becomes complicated in fundamentalist groups, where the membership to this community is reliant upon a strict belief in the literal interpretation of biblical texts.

**Table 3:** Statement of Belief for members of the Creation Research Society (Numbers 1992).

| Statement of Belief for CRS Members |
| --- |
| 1. The Bible is the written Word of God, and because it is inspired throughout, all its assertions are historically and scientifically true in all the original autographs. To the student of nature this means that the account of origins in Genesis is a factual presentation of simple historical truths. |
| 2. All basic types of living things, including man, were made by direct creative acts of God during the Creation Week described in Genesis. Whatever biological changes have occurred since Creation Week have accomplished only changes within the original created kinds. |
| 3. The great Flood described in Genesis, commonly referred to as the Noachian Flood, was an historic event worldwide in its extent and effect. |
| 4. We are an organization of Christian men of science who accept Jesus Christ as our Lord and Saviour. The account of the special creation of Adam and Eve as one man and woman and their subsequent fall into sin is the basis for our belief in the necessity of a Saviour for all mankind. Therefore, salvation can come only through accepting Jesus Christ as our Saviour. |

This prerequisite can be subtle and informal or explicit and official such as formal belief statements that act as prerequisites for membership to communities, such as Creation Research Society (Table 3).

While these formal statements are the most obvious clues to the demands of literal interpretation of Genesis for community membership, the informal variety are just as effective from an emotional perspective. The importance of community was a central point when I spoke to former members of these type of communities (See Conversations on Creation). When speaking to one of my in-laws who grew up in a young-earth creationist family, he described the difficulty he had in coming to terms with his doubts about the literal interpretation of Genesis because he knew that admitting these doubts to himself and others could cause him numerous friendships and strain his relationship to his family. Moreover, my conversations with another former creationist highlighted the fact that a sense of community and connection can be the driving force for individuals to pursue and maintain fundamentalist beliefs, while the absence of these relationships can be the impetus for a person to leave fundamentalist groups.

From an evolutionary standpoint the fact that creationists cling to their views in order to prevent their expulsion from a particular community, can be explained due to our ancestral tribal existence where community ties were essential for our survival. Now that we better understand the 'appeal' of religion in general and even how individuals can find psychological and social benefits through fundamentalist belief, we can better understand why creationists actively uphold their beliefs. These beliefs are the foundation of their meaning system, their terror management strategy and the key to many valued social connections. While it may appear as if creationists are being purposefully difficult by denying scientific data, we can now see that their behavior may actually be caused by a simple, innate desire to avoid pain and seek pleasure. The creationist version of this pain aversion | pleasure seeking behavior is the active maintenance of fundamentalist belief systems that ameliorate painful psychological states and increase pleasurable experiences by (1) creating meaning in a chaotic, unpredictable world, (2) offering them a perceived chance at immorality and (3) providing them with a key to membership in a coveted community.

# The Flipside of the Evolutionary Coin

*"In the distant future I see open fields for far more important researches. Psychology will be based on a new foundation, that of the necessary acquirement of each mental power and capacity by gradation. Light will be thrown on the origin of man and his history."*
                                                                    Charles Darwin (1859)

In the past section, we discussed the evolutionary causes for creationist *clinging* and established that from a psychological standpoint, it makes sense for creationists to maintain their fundamentalist views because they form the foundation of their meaning system, their world theory and self- theory. Yet, the flip side of this story is that the importance of these systems is also the reason why creationists actively *reject* evolution because they perceive evolution as a *threat* to these systems. Here it is important to remember that these meaning systems and terror management strategies are attributed to increased well-being and decreased anxiety as they provide a means of understanding and navigating an unpredictable environment and contribute to an alleviation of mortality anxiety. Thus, it is clear that a destabilization of these systems, theories and strategies could cause the opposite effect, i.e., increased anxiety and decreased well-being.

Similarly, as religious beliefs and even fundamentalist dogma act as the key to certain communities and thus social support, evolution poses a threat to these systems as it could cause the loss of certain social ties or even the excommunication from these communities (Reiss 2008). This is particularly true in the case of fundamentalism as evolution does directly contradict the literal interpretation of Genesis. While we briefly touched upon aversion and approach behavior in the last section, here we will concentrate on the most intense type of aversion behavior: the fight-or-flight response. Now that we understand the evolutionary causes for *clinging* to creationist beliefs and understand how evolution can be perceived as a threat to these systems, we can begin to comprehend how this physiological response can be activated when creationists are exposed to scientific data on evolution. Understanding the how and why of this system is essential at an educational level because once this system has been triggered, it is extremely difficult, if not impossible, for a student to cognitively engage with the data that are presented to them.

## *Fight-or-Flight Response to Lessons on Evolution*

In this section, we will primarily focus on the connection between learning about evolution and the destabilization of a creationist's concept of salvation in order to understand how learning about evolution can lead to a fight-or-flight response. While the belief in an afterlife may seem foreign to a many non-religious individuals and thus difficult to comprehend as a source of hope or anxiety, we have to realize that immortality is seen as a virtual certainty for many of our students; as professor of psychology Daniel M. Ogilvie states, "A majority of people worldwide believe that they and all other human beings are 'ensouled' and the majority of soul-believers endorse the idea that souls are released from the body when it ceases to function. For these people, the entrance of the soul into afterlife existence is taken to be a fact; a

**Table 4:** Common beliefs regarding salvation among all Christians according to Ogilvie.

| Basic Christian Beliefs on Salvation |
|---|
| People are born into a world of sin. The ritual of baptism is designed to cleanse the soul from inherited sin. But baptism alone does not guarantee salvation. |
| The only way to salvation is through Jesus Christ. |
| There will be a Day of Final Judgment. |
| The consequence of not being redeemed is eternal damnation. There are two deaths. One, the death of the body, is unavoidable. The second death can be averted by believing in Jesus and following his path to resurrection. Biblical interpreters vary in their visions of the second death. The mild form is being excluded from being in the presence of God. The harsh form is eternal damnation in the fiery pit of hell. |

life-sustaining, life-guiding, fear-reducing, mind-calming, unquestioned, reassuring, sometimes terrifying, 'it's going to happen', fact (2010)."

Interestingly, despite the split between the Catholic Church and the Protestant Church, the proliferation of protestant denominations and the multitude of differences and disagreements among Christian groups, there appears to be a set of ideas regarding salvation that all Christians agree upon, despite (Ogilvie 2010) (Table 4).

According to these beliefs, eternal life is dependent on adhering to particular religious customs such as baptism as well as lifestyle choices and belief systems, namely the acceptance of Jesus as the adherents Lord and Savior. According to these belief systems, failure to be baptized or failure to follow the rules set forth in the Bible could lead to eternal damnation. Thus, for many, a loss in this faith would mean that life would end when the physical body dies. Understandably, this image can cause severe anxiety. As I found out with in my conversations with former creationists, many of the teaching and theories taught in evangelical circles revolve around the necessity to prepare yourself for the 'end of times' where you will meet 'your maker'.

So, while an outsider may believe that a creationist is unable to 'grasp' modern science, in actuality they are not so troubled by the validity of the data, but instead much more concerned about the fate of their soul. Imagine a person who is completely convinced that eternal life will be granted to them through their faith and personal relationship to Jesus, but with the special caveat that this faith and relationship is based on a literal interpretation of Genesis, then suddenly it becomes clear that presenting this person with clear data on the organic origin of species is no longer a simple classroom lesson but instead a direct challenge to their belief system, their faith and ultimately, their eternal salvation.

The value of this afterlife is so high for some believers that it exceeds their value for their earthly lives leading to parasuicidal behavior such as the refusal of medical interventions despite imminent need (Tayler 2015). For non-believers it is thus possible to get a feeling for the importance of this idea of an afterlife by equating it with the value we hold for our earthly lives. Just as many of us try to extend and protect our earthly lives, by living carefully, healthily and thoughtfully, these austacious believers do all they can to retain their chances at eternal life by upholding their faith, and following the rules put forth to them by religious leaders. In comparison to other religious adherents who may obtain access to salvation simply by being a good person, the rules are much stricter in fundamentalist circles and thus the fear of losing one's salvation is much higher.

In essence, we can therefore understand the creationists' fear of evolution as a spiritual struggle for survival. Just as we are driven by evolutionary mechanisms to prolong our physical life, creationists appear to be driven by the same force to ensure their eternal life. In other words, while the creationists have accepted that worldly death is unavoidable, they still maintain hope in circumventing their second and ultimate death by securing access to the Kingdom of Heaven.

While the fight-or-flight response evolved to help us survive imminent, mortal danger, it can also be triggered in our modern lives through a number of non-life-threatening situations such as public speaking. Likewise, we see how this response can be elicited in the creationists through their perceived struggle for spiritual survival. In the next section we will dissect the biological mechanisms and evolutionary causes of the fight-or-flight response so that we better understand how this neurological system is triggered by intense fear responses, with the hope that we can also learn how to avoid triggering this system in and outside of the classroom.

## *Evolutionary Relics in the Brain*

To begin we will take a deep dive into our evolutionary history from a neurological perspective using the triune brain model as a guide, which was proposed by neuroscientist Paul D. MacLean in 1960. The triune brain model divides the brain into three main regions based on evolutionary history: the reptilian complex, the paleomammalian complex[16] (limbic system), and the neomammalian complex (neocortex), where the 'neocortex' represents the brain structures that are involved in advanced cognition; the 'limbic system' refers to the brain structures that are associated with social behaviors that arose during the age of mammals; and the 'reptilian complex' which refers to older brain structures that are related to the most basic of survival behaviors such as territoriality (MacLean 1990) (Figure 16).

While some modern neuroscientists disagree with the terminology of the limbic system and argue that the triune brain model is inaccurate due to its overgeneralized, its simplicity is also what makes the system so useful for non-specialist communications. While I agree that the triune model cannot be used to discuss the intricacies of neural functioning, it does serve our purpose of understanding how different brain structures affect our perceptions and our behavior.

Here we will concentrate on two main parts of this triune system: the limbic system and the neocortex with a particular focus on two structures within these systems, namely the amygdala in the limbic system and the prefrontal cortex in the neocortex as the interaction of these two structures plays an essential role in our emotional responses, emotional regulation, and our ability to think rationally. Understanding this is essential to developing better means of communicating effectively about science since the intellectual processing of scientific data requires that a person is functioning rationally and not out of affect.

---

[16]  Other names include: mammalian cortex or the paleopallium or intermediate brain, old mammalian brain.

## Triune Brain Model

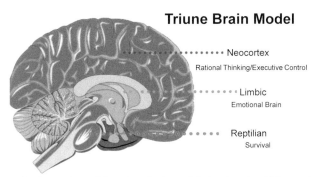

**Figure 16:** Generalized overview of the triune brain model showing the differentiation between the thinking, feeling and survival portions of the brain.

To begin with this analysis, let us look at the amygdala, which is located deep within the temporal lobes of the brain in all complex vertebrates, including humans. The amygdala plays a primary role in memory processing, decision-making and emotional responses, particularly the expression of fear, anxiety and aggression. The amygdala is also nicknamed the 'brain's watchdog' due to its role in a complex system that helps organisms recognize and react to perceived threats. While the amygdala's historical role in the evolution of mammals surely assisted in species survival, this system is maladapted for modern life as it responds to today's stresses (meeting a deadline) in the same way that it responded to mortal danger in the past. It is thus notoriously known for its tendency to overreact to environmental stimuli, as neuropsychologist Rick Hanson describes, "In the middle of your head, the central circuit of over reactivity has three parts to it: the amygdala, the hypothalamus, and the hippocampus. The almond-shaped amygdala does respond to positive events and feelings, but in most people, it is activated more by negative ones… . To initiate a fight-or-flight response, your amygdala sent alarm signals to your hypothalamus and to sympathetic nervous system control centers in your brain stem. Your hypothalamus sent out an urgent call for adrenaline, cortisol, norepinephrine, and other stress hormones. Now your heart was beating faster, your thoughts were speeding up, and you started feeling rattled or upset. Your hippocampus formed an initial neural trace of the experience—what happened, who said what, and how you felt—and then guided its consolidation in cortical memory networks so you could learn from it later. Connected by the neural equivalent of a four-lane superhighway, your activated amygdala commanded your hippocampus to prioritize this stressful experience for storage, even marking new baby neurons to be, in effect, forever fearful (Hanson 2013, p. 22)."

The speed and coordination of the limbic system is due to the fact that its main function is to initiate and manage the fight-or-flight response (MacLean 1990). The strength of these neural connections and thus the profound effect that this system has on our behavior is mostly likely due to the fact that this fear-driven survival system is older than the newer rational (cognitive) areas of the forebrain or neocortex. In fact, the hard wiring of the amygdala is so strong that it can even come to what is known as 'amygdala hijacking', where the amygdala can actually override *all* rational functioning in response to situations that are perceived as a threat. This sudden onset of strong emotions and impulsive, often aggressive behavior appears to act as a means for the brain to gear us up for a full fight-or-flight response.

What does this have to do with science education? As we have seen, the discussion of human origins is an emotional and sensitive topic for the creationist, one that is intrinsically connected to understanding themselves, the world around them and ultimately their chance at salvation. As evolution threatens to destabilize these systems, the creationist's mind can react to it in the same way that it would to physical danger to the body, i.e., by initiating a fight-or-flight response. When involved in a fight-or-flight response, a person is functioning on the basis of primitive neural structures and is unable to think rationally or process the data presented to them. In order for learning to occur, it is imperative that the activation of this fight-or-flight response is avoided so that the functioning of the prefrontal cortex, the area of the brain that controls reasoning, remains fully intact. As we will discuss in later portions of this book, the prefrontal cortex is vitally important for learning because it is responsible for our highest cognitive functions including self-awareness, social interactions, decision-making, planning and risk management (Teffer and Semendeferi 2012).

According to anthropologists Kate Teffer and Katerina Semendeferi, the "prefrontal cortex is critical to many cognitive abilities that are considered particularly human, and forms a large part of a large neural system crucial for normal socio-emotional and executive functioning in humans and other primates (2012, p. 191)." This is crucial when discussing evolution with creationists because conceptual change requires an active choice to engage with new material. As we discussed in the last chapter, a person must first see that their initial view is incompatible with newly presented scientific data and then *choose* to accept the new information while (partially or completely) rejecting their previous beliefs. As we will discuss in the next chapter, this process of assimilation and accommodation requires that a person has access to their highest cognitive powers, which is clearly not possible if their mind has been hijacked by the amygdala.

It is therefore important that we attempt to avoid triggering a fight-or-flight response and instead focus on activating other motivational systems that can support both better learning outcomes and improved social dynamics—namely our 'goal-seeking' system and our 'befriend' system. Both of these systems are also part of our neural circuity and also linked to our evolutionary past. The categorization of these motivation systems are also sometimes referred to as the avoid, approach and attach systems (Hanson 2011). Table 5 offers a brief neurological and evolutionary overview of the three major motivation systems: (1) threat system or fear-based, (2) drive system or goal-seeking system and (3) befriend system. These descriptions are based on Harvard psychologist Richard Siegel's book *The Science of Mindfulness: A Research-Based Path to Well-Being*.

Looking at this table, it should be clear that we want to be activating either the goal-seeking (approach) or tend-and-befriend (attach) systems, while avoiding the activation of the threat-system for two main reasons: (1) we want a system which has an activating effect, not inhibitory and one that (2) encourages positive emotional states. While all of these motivation systems have the potential to be activating systems, the threat-system also has an inhibitory function—the so-called 'freeze-or-

**Table 5:** The neurological and evolutionary overview of the three major motivation systems and how they affect cognition.

| Three Major Motivation Systems: Avoid, Approach and Attach | |
|---|---|
| **Motivation System** | **Emotional & Behavioral Response** |
| fight-or-flight system or threat system | <ul><li>The first system is focused on threats and is designed to help an individual escape a situation that is perceived as dangerous.</li><li>Physiology: involves the amygdala and the neurotransmitters: adrenaline (or epinephrine) and noradrenaline (or norepinephrine).</li><li>The neural system causes individuals to seek protection and safety. It can be activating or inhibiting—initiating an active fight-or-flight reaction or cause an individual to freeze or faint.</li><li>The activated feelings during a fight-or-flight reaction are often: anger, anxiety or even disgust.</li><li>When an individual is fearful of their own survival they will tune out the needs of others.</li></ul> |
| goal-seeking system or drive system | <ul><li>The second system involves our sense of drive, excitement, and even vitality. It is activated through our pursuit of pleasure or when we would like to achieve a particular goal.</li><li>Physiology: involves the nucleus accumbens (basal forebrain) and dopamine system.</li><li>This is an activating neural response, waking feelings such as interest and excitement.</li><li>Similar to the first system, when this system is activated, the individual is tuned into their own needs but not those of others.</li></ul> |
| tend-and-befriend system | <ul><li>The third system is particularly active in mammals, and is deeply involved in feelings of contentment, safety, and connection.</li><li>Physiology: involves oxytocin which is generated in the pituitary.</li><li>It's an affiliative system, meaning that it creates feelings of well-being.</li><li>This system is easily overridden by the threat system and to a lesser degree by the drive system. Yet when the other systems are quiet, the tend-and-befriend system flourishes allowing for feelings of compassion, connection and empathy.</li></ul> |

faint' side of the 'fight-or-flight' response. This inhibitory side of the threat-system is what is associated with cognitive shutdown. Moreover, the goal-seeking and tend-and-befriend systems are the only systems associated with positive emotions such as interest and excitement or well-being, while the threat-system is linked to negative emotional states such as anger, anxiety or even disgust. Obviously, when discussing the validity of evolution, we want to be fostering interest and excitement in science rather than anger, disgust or anxiety. Yet, here the neurological cards are stacked against us, as the threat system is ruled by the almighty amygdala and thus both the goal-seeking and tend-and-befriend system can easily be overridden by the threat-system.

It is possible to activate these systems using extrinsic motivational tactics either in the form of a 'stick' (fear-based) and a 'carrot' (reward-based) procedures. These external means of motivation work because they correspond to internal motivation systems, namely our most primal pleasure-seeking (approach) and pain-aversion (avoid) instincts. Due to the hard-wiring of the pain-aversion system, the 'stick' is often the more effective means of motivation. This is because from an evolutionarily standpoint, 'sticks' represent those objects or situations which could cause bodily harm or even death, such predators, fires, social aggression, etc., while 'carrots' represent pleasures that were also necessary for survival but did not carry the same urgency or impact, such as food, sex, social connection, etc. Because we are programmed for survival, 'sticks' receive a much greater neurological investment because it was necessary that ancient humans respond immediately to 'sticks' as failure to do otherwise could lead to sudden harm or even death (Hanson 2011). Although, the 'stick' is the more effective means of motivation, it is the less desirable tool in the classroom or general communication because this motivational system is associated with negative emotional states and thus has the potential to cause a cognitive shut down as part of the faint-or-freeze side of the fear response.

Unfortunately, the 'stick' remains a very common tool of motivation within fundamentalist communities, where members are continuously reminded of the eternal damnation that awaits those who do not follow the rules set forth by the community and the sacred texts. Although this may be effective in the short-run it can also be the reason why members eventually decide to leave such communities. As we will hear later in book, once doubt does arise in the mind of a fundamentalist, it quickly becomes an all-or-nothing situation. Because one is taught that doubting the word of God will lead to eternal damnation, a means of circumventing this is simply to become an atheist—while one gives up the chance at eternal life, it is thus also possible to avoid the threat and fear of Hell.

The difficulty of communication between science enthusiasts and creationists is further complicated by our brain's general negativity bias.[17] Not only is our brain wired to react more strongly to threats than opportunities, but the brain actually takes it one step farther and begins to perceive threats even where there are none present. One of the easiest ways to understand the evolutionary advantage of a negativity bias is the idea of a paper tiger. As Rick Hanson states, "Our ancestors could make two kinds of mistakes: (1) thinking there was a tiger in the bushes when there wasn't one, and (2) thinking there was no tiger in the bushes when there actually was one. The cost of the first mistake was needless anxiety, while the cost of the second one was death. Consequently, we evolved to make the first mistake a thousand times to avoid making the second mistake even once (2013, p. 23)."

Obviously, this negative neurological programming that primes us to recognize negative stimuli and remember negative experiences can have disastrous effects on our social lives, communication and education. Sadly, this paper tiger paranoia is also present in the science classroom. As we have discussed, the creationist students

---

[17] The extent of this negativity bias can be understood using the metaphor of glue and Teflon, where the brain acts like glue for negative stimuli and non-stick Teflon in response to positive stimuli—in other words, our brains are evolutionarily programmed to notice negative stimuli and remember negative situations to a much greater extent than it notices or remembers positive experiences (Hanson 2013).

come into the classroom already wary of evolution. They fear that their faith and thus, their chance at an afterlife, will be challenged and possibly lost. This fear is even more deeply rooted in those students who received lessons on the danger of evolution during their early childhood. As Hanson further explains, "There are even regimes in the amygdala specifically designed to prevent the unlearning of fear, especially from childhood experiences. As a result, we end up preoccupied by threats, that are actually smaller or more manageable than we'd feared, while overlooking opportunities that are actually greater than we'd hoped for. In effect, we've got a brain that is prone to 'paper tiger paranoia (Hanson 2013, p. 23)." This negativity bias means that the creationist's brain is extra vigilant and primed for a fight-or-flight response as soon as evolution is mentioned.

However, the creationists are not the only ones harboring an over-alert threat recognition system. Every time that we equate an individual creationist with the 'threat' of the creationist movement, we are much more likely to activate our own fight-or-flight response. While the creationist movement, if left unabated, could have devastating consequences on our educational system, the creationist student should be seen as a product of this movement and not the driver of it. It is therefore necessary that scientists and science educators guard against creating our own version of this paper tiger paranoia since it will have direct effects on the way in which we respond to creationists and their objections to evolution. If we perceive the creationist as a threat, our means of communication will most likely be aggressive or defensive in nature, whereby the creationists' misconceptions of scientists and science would be confirmed. If we are, however, able to recognize the creationist simply as a product of their upbringing and social context, space is created for an effective discourse where the creationists may develop the willingness and ability to hear our responses. Yet, developing this open stance and participating in an empathetic and tolerant discourse is often difficult due to the 'us' versus 'them' nature of the creationism | evolution conflict as we will discuss in the next section.

## Tribal Discussions Between Creationists and Scientists

The role of community and the essential human need for social connection plays into the creationism | evolution problem in two ways. First, as we discussed, in certain fundamentalist circles, a literal interpretation Genesis and a general belief in the infallibility of the Bible are necessary to be fully integrated into the community. For members of such communities, evolution is seen as a threat because it directly contradicts the literal interpretation of Genesis, and if one were to accept evolution or begin to question the literal interpretation of Genesis, they could become ostracized and subsequently lose a valued social support system. This recognition of evolution as a threat is exacerbated by the second component which is also a product of our evolutionary past, namely, the fact that our intrinsic need of and reliance upon community has also led to the development of behavioral tendencies that (1) allow us to quickly discriminate between those who belong to our community and those who do not, and (2) to treat those who do not belong as less valuable or even less human. This behavioral tendency is sometimes referred to as 'othering'. These two aspects often compound one another, leading to a clear 'us' versus 'them' situation, which we can observe raging around the debate over human origins with creationists comprising one 'tribe' and secularists the other.

Ironically, this added complexity to the creationism | evolution conundrum is also a product of our evolutionary past and this tendency to discriminate between in-group and out-group is hard-wired into our brain due to its ability to help primitive humans and other organisms survive in hostile conditions. In fact, the ability to identify who is an 'us' and who is a 'them' is so prevalent in the organic world that it can be found everywhere from the cellular level in the ability of our immune system to recognize self from non-self to complex political networks. As Konstantin Khalturin and Thomas Bosch stated, "All organisms rely on their capacity of self/nonself discrimination to rapidly detect approaching allogeneic cells as well as invading pathogenic microbes as foreign and to eliminate them. Failure to recognize nonself causes self-mating, germline parasitism and disease (2007, p. 4)."

When it comes to higher mammals or humans, this type of behavior is also sometimes referred to as kin discrimination, which is broadly defined as a change in social behavior that depends on the degree of genetic relatedness of the individuals. Beyond genetic relatedness our ability and tendency to discriminate between those 'who belongs to my group' is also hard-wired into our brains due to our ancestral need to quickly distinguish between 'us' and 'them' in order to protect family and other tribe members from physical threats and competition for resources with competing tribes. Here it is important to remember that within religious groups enduring family-like cooperation are created and maintained between non-family members (Steadman and Palmer 2008).

Yet, as useful as this ability may have been for hunters and gathers thousands of years ago, it has major psychological and societal (and as we will see—educational) disadvantages, as Susan Krauss Whitbourne states, "Perhaps there is some survival mechanism at work in formulating ingroup–outgroup distinctions. In our desire to feel safe, we bond together with those whom we see as most like us so that we can protect ourselves from those who might do us harm. The virtual fences we build keep the outsiders away and allow us to go on with our daily lives feeling protected and secure. However, it is precisely these fences that keep us from bonding with our fellow human beings and in this way, undercut our true security (2010)."

But what does this have to do with creationism | evolution conundrum? Here again we see how our inner Neanderthal is at play in the mental construction of 'tribes' consisting of creationists in one group and secularist in the other. This battle is obvious both inside and outside of the classroom, perpetuating fights that involve everything from textbook disclaimers to changes to state science standards as we discussed in detail in the chapter on legal conflicts. This 'us-versus-them' mentality is apparent not only in the proliferation of legal cases but also in the divisive speech, school board hearings, emotional debates, etc. This division is present in all forms of media and, as we saw in earlier chapters, has existed in the United States for over one hundred years. The result is that both sides feel threatened and are convinced that they are 'under attack'. These external societal and historical factors only act to inflate our sense of threat in the classroom, thereby increasing the likelihood of a fight-or-flight response to lessons on evolution.

Again, this is not only true for creationists, but also for secularists, scientists and educators who also feel like they are under attack. Looking back at previous chapters we see that there have been multiple attempts to undermine evolution by trying to have it prohibited, watered down or 'balanced' with religious doctrines (Table 1). For teachers, this sense of threat is compounded by the fact that think-tanks, such as the Discovery Institute, are currently focusing on developing more classroom-based interventions due to legal losses outlined earlier. These classroom-level strategies aimed at undermining evolution education put teachers at the center of the conflict. A clear example of this is the development of 'Ten Questions to Ask Your Biology Teacher About Evolution' by Jonathan Wells, which the Discovery Institute has also disseminated to parents and students in the form flyers with suggestions about how they can promote a 'healthy' in-class debate about the strengths and weakness of evolution (Figure 17). Just a few years after Wells came up with his set of Ten Questions, William Dembski followed suit, developing his 'Ten Questions to Ask Your Biology Teacher about Design' (Figure 18). These strategies are designed to and often succeed in disrupting classroom instruction and place the teacher in an uncomfortable position of feeling personally responsible for upholding the validity of the theory of evolution.

These type of 'attacks' make teaching that much more difficult, to the point that many teachers even resist teaching such polarizing topics because they have self-doubt about content knowledge and concerns about students' responses (Swim and Fraser 2013). So, it is not surprising that teaching evolution can at times be very stressful and that challenges presented by anti-evolution students may appear as a threat to our sensitive amygdala. Luckily, there are also a great number of organizations and individuals who support secular science education and have thus provided sample answers to all of Wells' questions, see for instance the answers provided by Panda's Thumb (https://pandasthumb.org/archives/2005/05/ten-questions-t-1.html, Accessed 29 September 2018) or by the National Center for Science Education (https://ncse.com/creationism/analysis/10-answers-to-jonathan-wellss-10-questions, Accessed 29 September 2018). Yet, as soon as pro-science organizations have addressed one threat, a new one appears and this cyclic nature of 'attack and respond' only further strengthens the sense of 'tribal' warfare.

So, while creationists perceive evolution and secularism as a threat to their worldview, world theory and self-theory, secularists also feel as if they are 'under attack' due to the fact that multiple organizations and individuals are actively developing materials to undermine quality science education. Yet, in the same way that we hope to avoid triggering the fight-or-flight response in our students, we must guard against the activation of our own misguided threat-recognition system. While it is vitally important that we vigorously oppose think-tanks and any legislation aimed at undermining quality science education, we need to maintain the ability to actively engage in conversations on evolution using our highest cognitive functions instead of simply reacting out of affect. We need to intellectually decouple the creationist student from the creationist movement because they are simply pawns in a much larger strategy orchestrated by national think-tanks. The utilization of our own higher cognitive functions (prefrontal cortex) will better allow us to identify the opportunity to communicate effectively instead of feeding into the propagated

## Ten Questions to Ask Your Biology Teacher About Evolution

#1 ORIGIN OF LIFE. Why do textbooks claim that the 1953 Miller-Urey experiment shows how life's building blocks may have formed on the early Earth — when conditions on the early Earth were probably nothing like those used in the experiment, and the origin of life remains a mystery?

#2 DARWIN'S TREE OF LIFE. Why don't textbooks discuss the "Cambrian explosion," in which all major animal groups appear together in the fossil record fully formed instead of branching from a common ancestor — thus contradicting the evolutionary tree of life?

#3 HOMOLOGY. Why do textbooks define homology as similarity due to common ancestry, then claim that it is evidence for common ancestry — a circular argument masquerading as scientific evidence?

#4 VERTEBRATE EMBRYOS. Why do textbooks use drawings of similarities in vertebrate embryos as evidence for their common ancestry — even though biologists have known for over a century that vertebrate embryos are not most similar in their early stages, and the drawings are faked?

#5 ARCHAEOPTERYX. Why do textbooks portray this fossil as the missing link between dinosaurs and modern birds — even though modern birds are probably not descended from it, and its supposed ancestors do not appear until millions of years after it?

#6 PEPPERED MOTHS. Why do textbooks use pictures of peppered moths camouflaged on tree trunks as evidence for natural selection — when biologists have known since the 1980s that the moths don't normally rest on tree trunks, and all the pictures have been staged?

#7 DARWIN'S FINCHES. Why do textbooks claim that beak changes in Galapagos finches during a severe drought can explain the origin of species by natural selection — even though the changes were reversed after the drought ended, and no net evolution occurred?

#8 MUTANT FRUIT FLIES. Why do textbooks use fruit flies with an extra pair of wings as evidence that DNA mutations can supply raw materials for evolution — even though the extra wings have no muscles and these disabled mutants cannot survive outside the laboratory?

#9 HUMAN ORIGINS. Why are artists' drawings of ape-like humans used to justify materialistic claims that we are just animals and our existence is a mere accident — when fossil experts cannot even agree on who our supposed ancestors were or what they looked like?

#10 EVOLUTION A FACT? Why are we told that Darwin's theory of evolution is a scientific fact — even though many of its claims are based on misrepresentations of the facts?

**Figure 17:** Ten Questions to Ask Your Biology Teacher about Evolution by Jonathan Wells as an example of classroom-based strategies aimed at disrupting lessons on evolution.

# Ten Questions to Ask Your Biology Teacher About Design

#1 DESIGN DETECTION. If nature, or some aspect of it, is intelligently designed, how can we tell?

#2 GENERALIZING SETI. The search for extraterrestrial intelligence (SETI) is a scientific research program that searches for signs of intelligence from distant space. Should biologists likewise search for signs of intelligence in biological systems? Why or why not?

#3 BIOLOGY'S INFORMATION PROBLEM. How do we account for the complex information-rich patterns in biological systems? Where did they originate?

#4 MOLECULAR MACHINES. Do any structures in the cell resemble machines designed by humans? How do we account for such structures?

#5 IRREDUCIBLE COMPLEXITY. What are irreducibly complex systems? Do such systems exist in biology? If so, are those systems evidence for design? If not, why not?

#6 REUSABLE PARTS. Human designers reuse designs that work well. Life forms also repeat the use of certain structures (the camera eye, for example). Is this evidence for common descent, evolutionary convergence, common design, or a combination of these?

#7 REVERSE ENGINEERING. In trying to understand biological systems, molecular biologists often need to "reverse engineer" them. Is this evidence that the systems were engineered to begin with?

#8 PREDICTIONS. Do intelligent design theory and neo-Darwinian theory make different predictions? Take, for instance, junk DNA. For which of the two theories would the idea that large stretches of DNA are junk be more plausible?

#9 FOLLOWING THE EVIDENCE. What evidence would convince you that intelligent design is true and neo-Darwinism is false? If no such evidence exists or indeed can exist, how can neo-Darwinism be a testable scientific theory?

#10 IDENTIFYING THE DESIGNER. Can we determine whether an object is designed without identifying or knowing anything about its designer? For instance, can we identify an object as an ancient artifact without knowing anything about the civilization that produced it?

**Figure 18:** Ten Questions to Ask Your Teacher About Design by William Dembski as an additional example of classroom-based strategies aimed at disrupting lessons on evolution.

controversy. Moreover, we may thereby counteract creationist misconceptions about the dangers of science (and scientists) and act as the impetus to the end of this cycle of attack and defend.

## Summary

For many it remains incomprehensible how masses of adult Americans continue to believe that the Earth is less than 10,000 years old despite clear scientific evidence to the contrary. As we have seen, though, these beliefs are not the product of ignorance or lack of access to correct information, but instead exist because creationists purposively and actively cling to a literal interpretation of the Bible, while simultaneously rejecting any scientific evidence that contradicts this interpretation. While this seems irrational to many, there appears to be clear evolutionary causes for creationists to act this way. Understanding our innate desire to avoid pain and seek pleasure allows us to better comprehend why a creationist would actively attempt to uphold their literalist beliefs in order to maintain their meaning system and terror management strategy which support them in their ability deal with difficult emotional states and cultivate positive emotional states. Moreover, the adherence to these fundamentalist beliefs also acts as the basis for valuable social connections, which offer further psychological and physical benefits. Thus, for the creationist student, the importance of their faith is psychologically and socially much more important to them than the fact that the literal interpretation of the biblical account of creation is negated by scientific data.

It is therefore important to realize that for creationist students, the presentation of scientific evidence is a direct challenge to deeply seeded belief systems. It is not surprising that creationists react negatively to evidence of evolution or remain unresponsive because the same reasons that underly creationist clinging are also the basis for them to reject evolution as it threatens to destabilize their world theories, self-theories and community membership. The fear of learning about evolution can become so strong that it results in a fight-or-flight response as the creationist is involved in a perceived struggle for spiritual survival and thereby entirely incapable of absorbing or processing any data presented to them. While this threat response is understandable at a neurological level and from an evolutionary perspective, it is also highly maladaptive as it overrides more beneficial motivation systems and hinders the person from achieving any real level of scientific literacy.

The beauty of the evolved brain, though, is that it is pliable, meaning that we can learn to recognize our maladapted survival drives and then develop the cognitive ability to actively and rationally choose how we want to respond in certain situations instead of being a slave to the overactive limbic system. The incredible truth is that we are capable of participating in our own furthered evolution due to the neuroplasticity of our brains, which can have profound effects not only on our personal well-being but also on our interpersonal skills. This has great implications in our modern lives, especially when we are dealing with emotionally salient topics such as evolution. This potential conscious evolution and the positive opportunities it presents for education and general communication will be the focus of the second part of the book.

# Part II
# Tend and Befriend

*Oxford Dictionary:*

*Tend: Care for or look after; give one's attention to. // Befriend: Act as or become a friend to (someone), especially when they are in need of help or support.*

*Wikipedia:*

*Tend-and-befriend is a behavior exhibited by some animals, including humans, in response to threat. It refers to protection of offspring (tending) and seeking out the social group for mutual defense (befriending). In evolutionary psychology, tend-and-befriend is theorized as having evolved as the typical female response to stress, just as the primary male response was fight-or-flight.*

# Chapter 5

# All in the Mind

While many publications have examined the history and presence of creationism, the main goal of this book is to understand creationist thought and the perpetual clinging to literalist beliefs at a cognitive level in order to find a means by which creationists can become more receptive to science. In order to develop the most effective educational and communication tools it is important to understand how learning occurs and how it is either supported or hindered by various mental and emotional processes. For that reason, we will take a moment to discuss some fundamental points that are necessary to understand in creationism and evolution at the level of the mind.

Specifically, we will define the idea of conceptual change in terms of science education in the first section, then look at how learning is affected by emotions in the second and finally we will take a brief glimpse at the unique landscape of the adolescent brain and how this relates to learning about evolution in high school.

## Conceptual Change

Conceptual change is the term used to describe the process by which one's perceptions or thought schemas transform over the course of a person's lifetime or throughout human development. It is a relevant and exciting topic not only in science education but also in the history and philosophy of science as well as cognitive psychology. From the perspective of cognitive psychology, the opposite of 'conceptual change' can range from 'closed mindedness' where individuals are reluctant to consider new ideas that may conflict with pre-established beliefs to 'belief perseverance' (also called 'conceptual conservatism') which is described as an individual's insistence on clinging to ideas that have already been refuted.

Within the framework of science education, the term 'conceptual change' has received a number of various definitions. According to Reinders Duit and David Treagust, the term 'conceptual change' is often misunderstood as 'an *exchange* of pre-instructional conceptions for the science concepts', yet they propose that 'conceptual change' should be understood as the fundamental *restructuring* of pre-instructional conceptual structures in order to allow for the acquisition of science

## Assimilation          Accommodation

New information
is assimilated to
fit pre-existing
conceptual
structures.

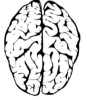

Pre-existing
conceptual
structures are
reworked to
accomodate new
information.

**Figure 19:** Overview of conceptual change in terms of Piaget's concepts of assimilation and accommodation.

concepts (2003). In this sense, conceptual change is much more a description of the pathway from the preconception to the acquisition of the intended knowledge (Duit 1999). The mechanisms of conceptual change in education are often described using Swiss developmental psychologist Jean Piaget's (1896–1980) notions of *assimilation* and *accommodation*, which were part of his larger cognitive development theory (Figure 19).

According to Piaget, cognitive development is a progressive restructuring of the brain which results through the combination of biological maturation and environmental stimulus. Piaget believed that as children mature they do not just learn about their environment in a passive way but instead develop certain schemas or patterns of knowledge in order to help them remember, organize and process information. According to Piaget's theory, when children encounter new information, they attempt to reconcile this new information with existing thought patterns (1985). He proposed that the human brain was programmed to seek a state of equilibrium, which it achieves educationally through the processes of 'assimilation' and 'accommodation' (Berger 2008). Piaget defines assimilation as the process by which new information is integrated into pre-existing cognitive schemes, i.e., this new information is interpreted to fit or *assimilate* to old ideas (Berger 2008). Accommodation, on the other hand, describes the means by which humans alter pre-existing cognitive schemes so that they become aligned with newly acquired information (Figure 19). In other words, assimilation processes use prior knowledge structures to understand new information while accommodation focuses more attention on the new information and requires a restructuring prior knowledge (Linnenbrink and Pintrich 2004).

In the case of a creationist learning about evolution, assimilation could be understood as a child who learns about the existence of dinosaur fossils (new information) and places them within the scope of a young Earth (existing thought schema), i.e., they accept that dinosaurs existed but assume that they existed just a couple thousand years ago along with human, likewise they might integrate the ideas of fossils into their belief in the great flood. Accommodation on the other hand could be understood as the same child learning about dinosaurs living millions of years ago (new information) and in order to integrate this new information, the person reworks their previous belief system thereby accepting that the Earth is ancient in order to accommodate enough time for the appearance and extinction of dinosaurs.

Both pathways of conceptual change are reliant upon a person's ability and willingness to cognitively engage with new material, while accommodation can be said to require an even higher degree of motivation and effort in order to restructure prior conceptual structures. This is true to an even greater extent when students are emotionally attached or invested in their pre-instructional conceptual structures, as in the case of creationist belief systems. Thus, getting a person to understand and accept evolution when they already maintain previous creationist beliefs requires a tremendous degree of willingness on the part of the creationist—which unfortunately is not often the case.

In the case of evolution and creationism, the processes of assimilation and accommodation are much easier when the student has an allegorical understanding of the Bible. This allegorical understanding means that the pre-instruction concepts are more pliable and thus it is easier to either assimilate knowledge of evolution into their ideas of creation or to alter these belief systems to accommodate for an understanding of how evolution is responsible for the organic origins of species. In the case of a literal interpretation of Genesis, assimilation is almost impossible as evolution requires vast expanses years of time, which is just simply not available in the concept of a young Earth. What this means is that the only option for a young-earth creationist to truly accept evolution is accommodation, which either takes place through the restructuring of their religious beliefs from a literal to an allegorical understanding of the Bible or in the most severe case that they begin to doubt the entire idea of special creation.

Yet, as we discussed in the previous section, there are many social, psychological and even evolutionary reasons for creationists to actively cling to their creationist beliefs, i.e., to engage in conceptual conservatism because accepting evolution would require them to entirely rework their understanding of themselves and their role in the world. This level of conceptual change cannot be demanded or expected to occur easily as it is contingent upon the person's willingness to engage with the scientific material and necessitates a high level of intrinsic motivation to reevaluate their previously held beliefs. Moreover, it is important to point out that this type of motivation usually requires a certain degree of dissatisfaction with previously held conceptual schemas, as Duit and Treagust stated, "If the learner was dissatisfied with his/her prior conception and an available replacement conception was intelligible, plausible and/or fruitful, accommodation of the new conception may follow (2003, p. 673)." Yet this is not often the case when creationists enter the classroom. While this puts educators at a disadvantage, adolescence may offer the best ontological period for this conceptual change to occur due to neurological changes that occur during this time span (this will be discussed in more detail later in *Evolution in the Mind of the Adolescent*).

As discussed in the previous chapter, creationism has been categorized as a misconception by many experts, while some argue that creationism should be understood as a worldview. As I stated, the combination of both assertions appears to be the most correct, as creationism is indeed a misconception of the natural world and simultaneously a worldview as it deeply affects the way in which the students perceive themselves and the meaning of their lives. Understanding this dual nature of creationist thought allows us to more easily recognize clear educational goals, while also acknowledging the emotional impact that the curriculum has on the student.

One of the major tasks of science education and science communication is to correct misunderstandings about natural processes and the natural world. Here it is clear that a belief in special creation (that animals appeared in approximately the same shape and form as present day) and a young Earth is incorrect and in direct conflict with data regarding biological evolution. While these misconceptions about the natural world can be corrected through effective communication, these same misconceptions are the lens through which creationists understand themselves and the world around them and thus by communicating about evolution we are also (unintentionally) destabilizing their sense of self and their place in the world. It is therefore not surprising that learning about evolution can be emotionally disturbing to these individuals. It also offers an explanation for why it is so difficult for these students to accept and understand the scientific principles of evolution as there appears to be an absolute lack of intrinsic motivation to engage with this newly presented information as it is counter to their worldview.

However, in order to increase a student's level of scientific literacy, they will need to be willing to relinquish their *fundamentalist* tendency to regard Genesis as the literal description of the origins of the organic world, because accepting the truth of evolution will mean that they come to terms with the fact that the Earth is ancient and that all animals and plants have evolved over millions of years and were not created in their present form. This is not a modest objective and the path to achieving this goal is neither simple nor easy. It must be clear that correcting the creationist misconception will also cause a shift in the way that a student sees themselves and the world around them. So, while there is a clear desire for science students to understand and accept biological evolution, legally and ethically this cannot be achieved by pressuring students to give up their faith.

Regardless of the lack of intention to challenge a student's religious beliefs, ultimately, this is what often occurs because when creationist students are confronted with evolution. In their eyes, they are presented with a choice of adhering to previously held creationist views regarding special creation (conceptual conservatism) or of finding a means to accept the scientific data on evolution (conceptual change). While many persons of faith are able to accommodate or at least assimilate evolution into their religious beliefs due to an allegorical understanding of Genesis, the creationist student will have expectedly more difficulty. From an educational standpoint, the creationist student has three basic options: They can (1) reject all scientific data that conflict with the literal interpretation of Genesis, (2) accept scientific data and reject the Genesis story entirely, or (3) they can accept the allegorical nature of the Genesis story and find a means to integrate the scientific data into their religious understanding of the world. More often than not, creationists are only aware of choices (1) and (2) and often lean towards choice (1) due to their emotional investment in their previous conceptual structures (Figure 12 and Figure 13).

In general, it can be said that emotions play a large role in conceptual change as studies on conceptual change have shown that the emotional response to a particular lesson influences the student's cognitive processes and this is most evident when students are confronted with scientific ideas that conflict with their own (Sinatra et al. 2014) as is the case with evolution.

Some experts such as Elizabeth Linnenbrink and Paul Pintrich propose that positive moods are beneficial for conceptual change as students who are in a good mood appear to be more likely to engage with anomalous information and are more open to new ideas, while students who experience negative emotional states such as anxiety are more likely to perceive new information as a threat (2004). According to the Cognitive Reconstruction of Knowledge Model (CRKM), students are much *less* likely to experience conceptual change if they have strong emotional ties to their prior beliefs (Dole and Sinatra 1998), which is clearly the case when it comes to individuals with long-held creationist beliefs. Because there is a large emotional investment in creationists beliefs and because these emotional ties hinder conceptual change, we will take a more in-depth look at the general role of emotions in learning and how positive emotional states can be fostered in the classroom to encourage better learning outcomes.

## Emotions and Learning

Emotions are not only present in the classroom but they have a significant impact on learning outcomes and play a crucial role in science learning (Alsop and Watts 2003; Laukenmann et al. 2003; Pekrun and Stephens 2012; Zembylas 2002, 2004, 2005). In fact, recent advances in the neuroscience of emotion have highlighted the connection between cognitive functions and affect, to the point that "evidence suggests that the aspects of cognition that we recruit most heavily in schools, namely learning, attention, memory, decision making, and social functioning, are both profoundly affected by and subsumed within the processes of emotion (Immordino-Yang and Damasio 2007, p. 3)". This information requires us to rethink the manner in which we teach and communicate about evolution in and outside of the classroom.

It has been shown that the effect of emotions in science learning is most evident when a person's ideas, worldviews or beliefs are challenged or are in conflict with the scientific data presented (Sinatra et al. 2014). In such cases negative emotions, such as frustration or anger, may foster resistance towards learning as the incoming scientific information is perceived as a threat (Gregoire 2003; Linnenbrink and Pintrich 2004). Most importantly, studies have shown emotions may support or impede conceptual change in science classrooms (Demastes-Southerland et al. 1995; Duit 1999; Duit and Treagust 2003). Thus, if we are interested in fostering scientific literacy, we must address the emotions surrounding the topic of human origins.

It is undeniable that the creationist student is primed to react negatively to lessons on evolution due to a battery of intrinsic and extrinsic factors—so how can we diminish these negative emotional states, or at least avoid aggravating these negative feelings? In search of possible solutions, we will first look at formal education settings to see how classroom environments affect the general emotional state of students and how positive classroom environments can support learning. The classroom as such can also be understood as a microcosm of society, meaning that what we learn about the dynamics of a classroom can be applied to general communication and learning within society at large.

While we cannot equate a positive classroom environment with the solution to internal emotional turmoil, we can see it as a means of not exacerbating the negative emotional state already present within the creationist student. This is important because research which has shown that students learn better when they are happy and feel and cared for (Noddings 2005), highlighting the link between positive social and emotional factors and positive academic outcomes. Teachers can actively promote such positive classroom environments by creating a place where students feel supported, cared for and encouraged to learn. In order to accomplish this, teachers need to develop a certain degree of social and emotional competence as studies have found that teachers' emotions have a profound influence on students' cognition, motivational levels and classroom behavior (Jennings and Greenberg 2011). Here Patricia Jennings and Mark Greenberg define socially and emotionally competent teachers as those teachers who have high degrees of self-awareness and social-awareness and are able to recognize their emotions and emotional patterns (2011). These teachers are better able to discern how student behavior affects their own internal environment and also understand how to generate positive emotions in themselves and in their students in order to motivate learning. These emotional and social skills have been shown to support a teacher's ability to build strong and trusting relationships with their students and to better navigate conflict situations (Greenberg et al. 2003a). This is crucial when teaching to mixed audiences as classroom disharmony can easily develop when emotions ride high (as is often the case with the topic of evolution).

It is easy to imagine how this emotional situation would be intensified when a student does not trust their teacher or feels like they are not accepted by their teacher. For that reason, it is important that educators remember that antagonizing a creationist student is the least likely way for the student to be able to understand and accept the theory of evolution. Moreover, it is for that same reason that I have highlighted the necessity for teachers and scientists to make a distinction between the creationist movement and the creationist student in order to avoid projecting our feelings of anxiety, frustration and anger onto the student.

I recently had this discussion with a colleague on the importance of creationist students feeling safe and supported in their learning environment (in order for them to learn about evolution) and he scoffed saying—"They feel safe enough to openly debate all material presented to them on evolution. How much more comfortable do you want them to feel?" For that reason, it is important that I point out here what I mean by 'safe'. Being argumentative and actively refusing to participate in learning is not a sign of feeling safe or at ease in a classroom. On the contrary, argumentative behavior is a sign of an agitated student who feels threatened and is most likely involved in a fight-or-flight response. This type of agitation and visceral aversion response renders the student incapable of learning and certainly incapable of restructuring their pre-instruction conceptual structures.

So how do we create a prosocial classroom environment or a safe place for discussion at a societal level? How do we help creationist students feel safe and secure so that they are willing to open their minds and become capable of learning?

Is this even really possible when what we are teaching is in direct conflict with their worldview? Can we correct their misconception about biological origins without toppling their psychological coping systems? The answers to these questions are complex but conceivable.

The answer to creating more social classrooms may lie in a combination of emotional awareness training and mindfulness practices (Jennings and Greenberg 2011). A study by Kemeny et al. tested the *Cultivating Emotional Balance* training program, which is a combination of Ekman's Emotion Awareness Training system and secularized mindfulness training. The program consists of 42 hours of training over a period of eight weeks and aims to reduce negative emotions while enhancing empathy and compassion. The program was tested using a randomized, controlled study involving 82 female teachers (pre-K to 12). When researchers followed up with the teachers after the program, the found that there was a significant decrease in the amount of self-reported depression and rumination and at the same time, the teachers reported an increased amount of emotional self-awareness (Kemeny et al. 2012).

Yet the effects of the program extended beyond the teachers' own emotional well-being and awareness. Although the training program only involved the teachers, follow up studies showed that teachers' newly acquired emotional competencies had many observable positive effects on the general classroom environment (Jennings 2007). This extended effect on the entire classroom dynamic is most likely due to the fact that the intervention teachers displayed an increase in compassionate responses to suffering during experimental tasks (Kemeny et al. 2012).

These emotionally competent teachers were not only more compassionate and empathetic, but they also developed a better understanding of how they could use positive emotions such as joy and enthusiasm to motivate students to learn (Jennings and Greenberg 2011). Emotionally competent teachers are thus better able to identify emotions in others and recognize how their emotional states affect those around them (Hen and Sharabi-Nov 2014). These capabilities then allow such teachers to create and maintain supportive relationships with their colleagues and students and support them in finding solutions during emotionally-difficult situations (Greenberg et al. 2003b).

The ability for a teacher to deal with students in stressful situations is crucial when it comes to discussing resolutions to the creationism | evolution conflict. This battle has been raging in American schools for over one hundred years and while the conflict between creationism and evolution may be purely intellectual for many secularists, we have seen that the issues run much deeper for the creationist, for whom this is an emotional, psychological, spiritual and social issue. This differentiation explains why we have not yet been successful due to the fact that the majority of previous attempts to address this problem have relied on a purely intellectual approach to the problem, using facts, data, diagrams and reasoning to convince the creationist. The emotional side of this issue has been neglected for far too long. While many educators and scientists wish that the creationists would move in a swift and direct line from denial of evolution to acceptance of evolution, that is simply unrealistic. If we return for a moment to the idea of the parallels between grief and creationists' confrontation with evolution, we can see that the line between denial and acceptance inevitably runs through a series of messy, emotional states (Figure 15).

The discussion of human origins is inherently psychological for the creationist and it is therefore naïve to believe that they can jump from denial to acceptance without an emotional response. As we will hear from personal accounts later in this book, "The personal and emotional elements behind these beliefs are, in my opinion, what really drives creationism. Yet they have never been addressed in any debates that I've seen. Creationism is an emotional battle not a logical one (Kron 2018)." This intrinsically emotional context of creationist beliefs means that we will need to expect and accept this emotional response as a necessary component of their path from the rejection of evolution to the acceptance of evolution as a scientific fact.

Furthermore, we cannot expect this change to occur over night. Lessons on evolution, whether in the classroom or as part of general science communication, are just the beginning of the process of conceptual change. In a way, these lessons can be seen as simple seeds that may require many years to root. While we may not be able to ensure that students move from denial of evolution to acceptance of evolution within a given school year or in response to a particular lesson, we can provide conditions, such as a positive learning environment and a trusting relationship to a teacher, that make it much more likely for that seed to take root someday.

While discussing the role of teachers' emotional competencies and positive classroom environment may not seem immediately relevant to teaching science or increasing scientific literacy, we can now see why it is necessary to address emotions and emotional climates due to the profound role that emotions play in the creationism | evolution conflict. But what about the students? We have spent so much time in this section focusing on the teachers and while they play a central role, they are still limited in their efficacy due to the inherent two-way nature of human relationships. For that reason, we will devote the next section to looking inside the mind of the high school student to see how the emotional nature of this topic is further magnified due to a restructuring of the brain that occurs during adolescence.

## Evolution in the Mind of the Adolescent

Evolution is taught most extensively in high school, meaning that evolution is most commonly taught to individuals who have not yet reached adulthood and whose minds are still in the process of developing. While teaching evolution to adolescent students presents certain disadvantages, there are also a number of clear benefits of teaching to this age group. While adolescents are notoriously known for being moody, hormonal, unpredictable and emotional. These over generalizations, stereotypes or myths can be limiting and disingenuous as clinical professor of psychiatry Dan Siegel states, "The adolescent period is amazing. People often view it in a negative light, thinking that adolescents are 'immature' or 'are going to lose their minds' or that 'they have raging hormones'. These are myths that mislead us and actually disempower us (2015a)."

The truth of the matter is that while adolescents do differ in the way they behave and make decisions, the neurological reasons that cause these differences present a great opportunity for conceptual change. From an educational stand-point, the potential embedded in these neurological differences might in fact outweigh the pitfalls. To begin we will look at the potential difficulty of teaching students at this age and then move on to the positive changes that occur at a neurological level during adolescence and how even some of the 'difficulties' may present learning opportunities, especially with regard to creationist students learning about evolution.

Our knowledge about the developing brain only continues to increase as technology advances, particularly through the use of magnetic resonance imaging (MRI) and functional MRI studies. It is now possible to scan the living human brain and see how it changes—both in its structure and organization—throughout a person's lifespan. While scientists and psychologists worldwide have been aware of the quick and dramatic developments of the brain throughout early childhood, scientists are just now becoming aware of the second major brain growth and restructuring period that takes place during adolescence.

Thanks to emerging research, it is now possible to explain some of the typically 'difficult' adolescent behavior on the basis of the developmental relationship between the amygdala and the prefrontal cortex during adolescence. While we discussed the role of the amygdala and prefrontal cortex briefly in the past chapter with regard to emotional response and rational thinking, we will now look in more depth at how these two structures change and develop during adolescence and why this is relevant for high school science education.

According to the American Academy of Child and Adolescent Psychiatry, the amygdala is fully (or overly) developed during adolescence, while the prefrontal cortex, responsible for rational thought, is not fully developed until early adulthood and is in fact undergoes major reconstruction during adolescence (2016). Thus, one of the proposed reasons for why adolescent behavior differs so greatly from adult behavior is due to the temporal differences in the developmental relationship between the amygdala and the prefrontal cortex during this stage of development.

Due to the imbalance between the developmental age of the amygdala and prefrontal cortex, adolescents differ from adults not only in their behavior but also in their emotional sensitivity. Although it is difficult to generalize brain functions, research conducted by Dutch scientists at Leiden University has shown that the amygdala appears to be more active in adolescent brains than in adult brains and that this overactive amygdala helps explain why teens' negative emotions, such as fear or sadness, may be experienced more intensely (Pannekoek et al. 2014). Research has also found that an increase in the volume of amygdala during adolescence can also result in a predisposition toward negative emotional states that can interfere with cognitive functioning and behavioral regulation (Teffer and Semendeferi 2012).

At the same time, the prefrontal cortex actually *loses* grey matter as it undergoes a striking and prolonged amount of structural changes. This loss of grey matter is the result of the synaptic 'pruning' which is part of a specialization process of brain. According to Sarah-Jayne Blakemore, professor of cognitive neuroscience, this *decrease* in grey matter volume begins during adolescence and continues to decline until into the early twenties (2014), meaning that this neurological transformation is also taking place in the minds of undergraduate students. The fact that this part of the brain is still maturing *and* undergoing major restructuring well into adulthood means that adolescents are not only prone to feeling emotions more intensely (particularly negative feelings) but that they also lack a true means of processing emotions and managing their emotional reactions.

Adolescence is a period marked by emotional turbulence as the amygdala guides 'gut' reactions to situations instead of clear rational thought or planning (Yurgelun-Todd 2002). In general, adolescents rely on older, cruder areas of the brain and it is thus much more likely for a teenager to be the victim of an amygdala hijacking than an adult. This emotionally turbulent period presents clear hurdles when teachers (or parents) expect high school students to act as rational as their adult counterparts. With regards to the evolution | creationism conflict, this means that once a negative emotion has been elicited in the creationist student, it is much more likely to be felt even more intensely and these students will have even more difficulty regulating this negative emotion than an adult would, meaning that learning is hindered to an even greater extent. Yet, if we can help keep negative emotional states to a minimum, the changing organization of the adolescent brain does have a number of major positive sides and offers a number of unique learning opportunities that are particularly helpful with regard to increasing a students' willingness to engage with new material and their ability to restructure previously held conceptual schemas.

As we discussed, the emotional (even irrational) life of teenagers is largely due to a restructuring and pruning of the prefrontal cortex. While this does present difficulties for the students and teachers alike, due to the increased likelihood of an amygdala hijacking, there is an upside. The reason why the prefrontal cortex loses grey matter during adolescence is because the brain is undergoing a major reconstruction, in essence it is being remodeled as it becomes more specialized. As the brain becomes more specialized, the brain is 'pruned', meaning that pathways that are not frequently used are removed. At the same time, a part of this neurological remodeling involves a rapid increase in connections between brain cells that make those pathways more effective and more coordinated through increased myelination during this time which subsequently leads to more coordinated thought, action and behavior (Siegel 2015b). This remodeling is the quintessence of neuroplasticity, known in day-to-day life by the concept of 'use it or lose it' or 'nerves that fire together wire together'. Why is this good news in terms of science education?

First of all, as Blakemore points out, this synaptic pruning fine tunes the brain to better fit its environment (2014) and Siegel goes so far as to say that this remodeling process can be actively guided by practicing skills (and characteristics) during this time that we wish to strengthen (2015a). What this means in terms of science education is that skill sets and thinking patterns that are taught and practiced during this developmental period can potentially be incorporated into this specialization process. In other words, teaching critical thinking skills and scientific investigation skills during this sensitive period of brain development could result in these processes being hard-wired in the brain's architecture, thus making it far more likely that these skills are retained and used in the future. At the same time, if creationist teenagers abstain even temporarily from practicing fundamentalist thought patterns, i.e., not relying 'blindly' on sacred texts as the foundation of their understanding of the natural world, then it is possible that these thought patterns become pruned, thus decreasing this behavioral tendency.

Some of the typical 'pitfalls' of teenage years such as their willingness to take risks and their increased interest in obtaining the approval of their peers can even be advantageous for education. While these behavioral traits are often seen in a negative light as they are associated with dangerous or unhealthy behaviors, such as risky driving or drug use, they are actually part of the natural process of becoming progressively more independent from their parents (Blakemore 2014). Although the concept of peer-approval is often associated with negative 'peer-pressure', it can also present a positive opportunity for students to turn away from fundamentalist ideologies in favor of scientific explanations as a means of gaining acceptance among their peers. As Linda S. Behar-Horenstein, distinguished teaching scholar and professor of education, points out, "As these students become adolescents, the need for approval changes from teacher-based to peer-based as the students' reference group changes. Adolescents are motivated by social goals as much if not more than academic goals so that the adolescent culture creates their cognition. Adolescents' need for inclusion and avoidance of exclusion motivates their behavior toward school, activities, and learning. These principles dramatically alter how teachers motivate adolescents versus motivating younger students (2006, pp. 310–311)". Thus, we can see how a creationist student would be motivated to learn about evolution when their peer-group accepts evolutionary theory. Yet when the creationist students' only peer-group are other literalist believers of Genesis who also deny evolution, then there will only be a higher level of motivation to maintain these fundamentalist beliefs.

Irrespective of the peer-groups, 'risky' and 'rebellious' adolescent behavior in itself may also represent an opportunity for intellectual emancipation as students become more willing to question the authority of the dogma and the subsequent belief systems that have been passed down to them from 'authoritative' figures throughout their childhood. This ability to question or doubt previously formed thought schemas is a prerequisite for conceptual change, particularly in the case of creationism. As we discussed in the previous section, for conceptual change to occur, an individual's willingness to restructure pre-instructional concepts often necessitates a certain degree of dissatisfaction with previously held beliefs (Duit and Treagust 2003). The necessity of this dissatisfaction is ultimately why adolescence may present the golden opportunity to teach creationist students about evolution as they may become dissatisfied with the restrictions imposed by fundamentalist thinking and rules as part of their movement towards becoming autonomous.

## Summary

Conceptual change is the process by which students are able to restructure pre-instructional thought schemas in order to understand, process and integrate newly presented information. For our purposes, creationist ideologies represent a pre-instructional conceptual structure, while evolution would be the newly introduced information. The connection between creationist ideologies and the creationist's self-theory often leads to a situation where a student displays 'conceptual conservatism' and is unwilling to engage with new material that contradicts previously held beliefs. Conceptual change often presupposes a certain degree of dissatisfaction with prior conceptual structures as the process is reliant upon a person's willingness to cognitively engage in new material, while altering previously held thought schemas.

This obviously presents science teachers and science communicators with a difficult situation because a large degree of restructuring of these conceptual structures is necessary for a student to move from a literal interpretation of Genesis to an understanding and acceptance of an ancient Earth and organic evolution. While getting a person to understand and accept evolution when they already maintain creationist beliefs may seem close to impossible, we have seen that adolescence offers a unique neurological opportunity due to the restructuring of the prefrontal cortex during this time.

The restructuring and pruning of the prefrontal cortex during adolescence means that it is possible to uproot old thought patterns and create new thought patterns simply by practicing certain thought habits while abstaining from others. In other words, if creationist students can be encouraged to think critically and analytically in the classroom, this may offset their propensity to rely on sacred texts blindly for their understanding of the world around them. Likewise, the development of autonomy during adolescence could cause creationist students to become more willing to question or even challenge authority figures and ideas passed down by those authority figures. This type of questioning of fundamentalist dogma can be strengthened through a teenager's desire to gain acceptance within a more secular peer-group. Of course, all of this is reliant upon the existence of opportune situations that may or may not come to fruition but as discussed, the development of pro-social classrooms that are associated with increase positive affect, could provide fertile ground for seeds of scientific literacy to be planted and later take root.

# Chapter 6
# Conscious Evolution

In the chapter *Evolutionary Causes of Creationist Clinging* we took a look at our evolutionary past and how our primitive past plays a role in the perpetuation of creationism due to our innate desire to avoid pain, seek pleasure and belong to a community. We have seen how our tribal past also exacerbates the creationism | evolution conflict as we are primed to identify those who do not belong to our community as potential threats. While our evolutionary history has provided us with a powerful survival framework, this framework can be disadvantageous in our modern world as the same strategies that helped our ancestors survive in primitive conditions now lead to unnecessary division in modern societies as well as unnecessary conflict and suffering. Without even being consciously aware of it, we react to one another based on old evolutionary tendencies. Yet, by understanding how our biological history has predisposed us to certain behaviors, we also gain insight into possible solutions.

In the same way that I decided to approach my understanding of creationism from a biological stand point in order to discern whether there were any evolutionary causes for creationist clinging, I have also attempted to seek a solution to this conflict by turning to science—in particular our understanding of neurobiology. Based on what we have learned about the role of various neurological structures and their effect on our perception and reaction to the world, I am certain that 'belief perseverance' can be resolved through increased understanding and implementation of the research currently being conducted in the fields of neurobiology, cognitive science and developmental psychology. One of the most promising finding from these fields is that the neuroplastic nature of our brains offers seemingly limitless opportunity to alter the ways that we perceive and engage with ourselves and the world around us. What this means is that we have the potential to consciously and intentionally guide mental processes and development, which allows for greater freedom and control over our own perceptions, learning and behavior. This understanding of neuroplasticity opens the doors to our ability to cognitively engage in the progressive evolution of our minds, thereby creating new opportunities for learning, conceptual change and societal engagement.

## Mindfulness and Neural Plasticity

As early as 1949, Donald Hebb proposed the basic mechanism of synaptic plasticity as an explanation for how neurological adaptations occur during the learning process. While his theory is often referred to as Hebbian theory, Hebb's rule, Hebb's postulate or cell assembly theory in scientific discourse, it is most well-known in popular literature through its sync summary, "Neurons that fire together wire together", offered in 1992 by Siegrid Lowel and Wolf Singer in their paper *Selection of intrinsic horizontal connections in the visual cortex by correlated neuronal activity*— although it should be noted that the original sentence was "neurons wire together if they fire together (p. 211)." While this insight into synaptic plasticity allowed scientists to understand how repeated actions led to ever better and more automated physical movements, such as a golf swing, it also led to the recognition that certain thought patterns and behavioral patterns are also 'hard-wired' through repetition. In this way, it was found that negative thinking patterns and rumination, for instance, were deeply connected to depression. This realization led to the development of new psychotherapeutic interventions that aimed to disrupt this thought-looping in order to allow the synapses to form new pathways. The most exciting research with regard to the topic at hand is that data have shown that humans can actively influence how the brain develops new synapses through contemplative practices, mindfulness practices, cognitive behavioral therapy, etc. These practices essentially allow us to intervene in our own conscious evolution. Neuropsychologist Rick Hanson refers to this process as self-directed neuroplasticity, where individuals are able to reconstruct the structure and organization of their brain through contemplative practices (Figure 20). According to Hanson, we can "use the mind to change the brain to change the mind for the better (2011, p. 12)."

According to Center for Contemplative Mind in Society (CMind), "Contemplative practices are practical, radical, and transformative, developing capacities for deep concentration and quieting the mind in the midst of the action and distraction that fills everyday life. This state of calm centeredness is an aid to exploration of meaning, purpose and values. Contemplative practices can help develop greater empathy

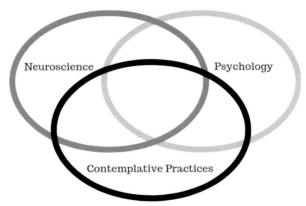

**Figure 20:** Hanson's idea of self-directed neuroplasticity using contemplative practices to influence mental and neurological processes.

and communication skills, improve focus and attention, reduce stress and enhance creativity, supporting a loving and compassionate approach to life (CMind 2015)." Clearly this type of development could be very useful in ameliorating the negative emotions involved in the creationism | evolution conflict. While contemplative practices can take on many forms—from journaling to pilgrimages—we will focus on mindfulness practices as a form of contemplative practice because there is a wealth of research and data available that specifically address its usefulness for non-pathological purposes due to its effectiveness in changing thought patterns and decreasing emotional over reactivity. In fact, there has been a surge of publications on the applicability of mindfulness since 2011 and it is estimated that over 200 papers are published per year on the topic of mindfulness (Tabak et al. 2015).

Mindfulness practice involves the cultivation of undistracted attentiveness to what is being experienced in the present moment (Fronsdal 1998). Mindfulness practice thus allows us to clearly recognize what is going on within us during a given situation and to respond to the situation consciously instead of being driven by emotional reactivity. This recognition and awareness in turn act as a means of letting go of unskillful thought and behavioral patterns. Mindfulness can thus support the development of self-regulatory skills associated with the ability to be aware of emotions and consciously choose how to react to those emotions. It is associated with the ability to observe without judgement and to live fully within the present moment.

Because mindfulness practices have the potential to guide self-directed neuroplastic changes, it presents itself as a unique tool in helping to resolve the creationism | evolution conundrum. As we discussed in the last chapter, research has actually shown that a combination of emotional awareness training and mindfulness practices may offer the key to creating more social classrooms (Jennings and Greenberg 2011). These types of positive learning environments allow students to develop more trust in their teacher and make meaningful connections to their peers. This educational safe-zone in turn leads to an increase in a student's willingness to engage with new data, even when it does not fit in the pre-existing thought schema.

Moreover, mindfulness can also assist persons on both sides of the creationism | evolution debate to recognize when their perception and reaction to the world around them is being driven by maladaptive, primitive reactions instead of purposeful thought and rationality. Mindfulness can help the creationists develop the ability to become aware of their clinging to fundamentalist teachings and recognize how this clinging affects not only their own internal environment but also their ability to engage with the world around them. There are also a number of benefits for secularists and even 'new atheists' who can better identify when they are being driven by prejudicial judgments, which may cause them to react to creationists in a manner that only encourages creationists to grasp more tightly to fundamentalist ideologies and reject science. Most importantly, this new awareness of thought patterns and reactivity enables practitioners to develop a higher degree of cognitive flexibility which in turn allows them to change these habits.

Lastly, as we will see, mindfulness is also a profoundly useful tool in learning how to deal with difficult emotional states. As we have already established, fundamentalist beliefs can form the foundation of an individual's self-theory and world theory and that both of these can become destabilized through the questioning of central ideologies such as the literal interpretation of the Bible. While it may not be a teacher's intention to cause a student to give up their religious views, this may be an inadvertent side-effect of increased scientific literacy. This is especially true when these systems are based on a literal interpretation of Genesis as this prevents a student from accommodating evolution into this conceptual structure, meaning that the acceptance of evolution would require a complete restructuring or loss of this thought schema. While the loss of fundamentalist thought patterns is beneficial to both the student and the teacher, it can also cause a subsequent destabilization of the students' meaning system and terror management strategy, placing that student in a particularly susceptible position in terms of emotionality at an age when emotions are already running high. It is therefore additionally useful to teach mindfulness practices to this age group so that they have an alternative means of dealing with negative emotional states. In this way mindfulness practice can act as a set of coping skills that could offset the disruption of previous coping mechanisms, i.e., relying on fundamentalist principles to make sense of their world.

Because mindfulness practitioners do develop a greater ability to deal with difficult emotional states and undo unhealthy thought patterns, it is not surprising that these types of practices have received a great deal of attention from professionals from the fields of medicine and psychology. Mindfulness was first successfully implemented within the healthcare branch in 1979 and numerous studies since then continue to substantiate its efficacy in reducing stress, anxiety, depression, etc. The following table (Table 6) offers an overview of current mindfulness-based mental healthcare programs as summarized by neuropsychologist Ronald Siegel (2014).

As can be seen from the table, mindfulness-based programs have proven to be effective in treating multiple, severe physical and psychotic conditions. We can see that Mindfulness-based cognitive therapy (MBCT) reflects Hanson's idea of self-directed neuroplasticity and my idea of conscious evolution. While we do share an evolutionary history with all living beings, we are truly different from all other animals in our cognitive abilities and adaptability, which enable us to override maladaptive reactionary behavioral tendencies in favor of higher cognitive responses. This is accomplished by engaging in our own conscious evolution or in the words of MBCT, by changing our thought patterns using behavioral learning principles.

Here, I should make it very clear that I am not proposing that creationists should be seen or treated like patients, nor am I insinuating that creationist thought is a type of psychosis. What I am trying to point out is that mindfulness has been effective in treating major psychological ailments such as borderline personality disorders, schizophrenia and drug addiction because it acts as a tool that patients can use to better deal with intense, difficult emotions and to disrupt those thinking patterns that cause or perpetuate these negative emotional states.

How mindfulness is defined in the medical world differs from the general definition of mindfulness. Bishop et al. have attempted to create an operational definition of mindfulness and propose the use of a two-component model of mindfulness, where "The first component involves the self-regulation of attention so that it is maintained on immediate experience, thereby allowing for increased recognition of mental events in the present moment. The second component involves

**Table 6:** Background, overview and effectiveness of mindfulness-based healthcare programs (Siegel 2014).

| Overview of Current Mindfulness Programs | |
|---|---|
| **Program types** | **Purposes and goals** |
| Mindfulness-based stress reduction (MBSR) | • Started in 1979 by Jon Kabat-Zinn, who is responsible for the wide-scale adoption of mindfulness practices in many different branches of healthcare.<br>• Aim to treat physical pain and stress-related disorders.<br>• Most widely studied and used mindfulness-based program<br>• Major component of MBSR is mindfulness meditation. In addition to the informal and formal meditation practiced, it can also be combined with body scan visualizations and hatha yoga sequences.<br>• MBSR has been shown to improve pain ratings, lower anxiety, decrease the frequency of panic disorders, alleviate binge eating and fibromyalgia symptoms, as well as reduce mood disturbances and stress levels during cancer treatments, as well as assist patients in dealing with a variety of other medical conditions and general psychological symptoms. |
| Mindfulness-based cognitive therapy (MBCT) | • Mindfulness-based cognitive therapy is based on MBSR and belongs to the third wave of behavior therapies.[18]<br>• MBCT is based on the idea that humans are very different from animals in that we are thinking-beings and are capable of changing our thought patterns using behavioral learning principles.<br>• Carl Rogers emphasized acceptance as a precondition of change. Meaning that no change in thought patterns or behaviors can occur without profound acceptance of the experience. This is thus the starting point for all mindfulness- and acceptance-based treatments listed here.<br>• MBCT has been proven to have dramatic outcomes, as clinical studies have shown for example that after one year, two-thirds of the depression patients who had participated in at least four sessions of MBCT remained depression free, while only one-third of the people who had the usual treatment remained depression-free. |
| Dialectical behavior therapy (DBT) | • Developed by Marsha Linehan to treat severe emotional disregulation, as seen in borderline patients<br>• This program combines mindfulness practice, particularly in the Zen tradition, and cognitive behavior therapy in an attempt to teach patients about the role of acceptance and change in difficult situations.<br>• Many studies have shown the efficacy of this program with regards to reduction in the frequency of parasuicidal and self-injurious behavior, as well as a decrease in drug use and eating disorders.<br>• Moreover, it has been shown that the techniques used in DBT are also functional for individuals without an emotional disregulation, in that it also supports them in developing more self-acceptance and the ability to deal with emotional situations and assist them in changing harmful behavioral patterns. |

*Table 6 contd. ...*

---

[18]  The first wave was initiated by Ivan Pavlov's discovery of unconditional stimulus, the second wave was started by Burrhus Frederic Skinner's studies of operant conditioning, and Carl Rogers' teachings brought about the third.

... *Table 6 contd.*

| Overview of Current Mindfulness Programs | |
|---|---|
| **Program types** | **Purposes and goals** |
| Acceptance and commitment therapy (ACT) | ▪ Developed from the 1980s to late 1990s by Steve Hayes and his colleagues.<br>▪ This is a mindfulness-, acceptance- and value-based psychotherapy program that integrates a number of Eastern meditative traditions yet is presented in a manner more accessible for Western practitioners.<br>▪ ACT programs have been used as a part of individual therapy and increasingly in group settings.<br>▪ ACT programs have been effective in reducing social anxiety, disability due to pain and rehospitalization of psychotic patients. It has also been shown to be more effective than nicotine replacement therapies in attempts to quit smoking.<br>▪ ACT incorporates several components: creative hopelessness, cognitive defusion, or deliteralization, acceptance, and self as context, which is about identifying with the observer of the thoughts, not the thinker.<br>▪ This last component focuses on 'valuing', which involves redirecting one's life to what gives it meaning. |

adopting a particular orientation toward one's experiences in the present moment, an orientation that is characterized by curiosity, openness, and acceptance (2004, p. 232)." While other definitions exist, the emphasis remains on the fact that mindfulness is strongly linked to increased self-regulation due to general increases in self-awareness coupled with increased abilities to maintain attention and regulate emotion (Tang et al. 2015).

For that reason, it is not surprising that many studies have provided evidence that mindfulness-based practices have great potential beyond addressing pathogenic issues such as attention training, emotional regulation, stress management, social and emotional learning, compassion, resilience and wellness (Sheinman and Hadar 2016). These points are very relevant to lessons and discussions involving evolution because mindfulness is an effective tool that allows practitioners to monitor and ease emotional states in order to support and enhance higher cognitive functions. In the next section we will take a particular look at the application of mindfulness practices in addressing the emotional complexity of 'controversial' topics such as evolution.

## *Mindfulness' Effects on the Brain*

As the interest in mindfulness grows, so does the body of research on the effects of mindfulness on our bodies and brains. Recent research using brain imaging techniques has revealed that mindfulness practices can profoundly change not only the manner in which we think, but also the means by which different regions of the brain communicate with each other. Even more interesting is that scientists believe that these changes can become permanent (Ireland 2014). Most exciting for our current discussion is the fact that mindfulness practices appear to have a major effect on the amygdala and its relationship to the rest of the brain. As we discussed in earlier chapters, the amygdala is part of the limbic part of the brain and acts as the brain's watch dog, while the prefrontal cortex belongs to the neocortex and could be considered the computer of the brain that guides rational, reflective behavior. Getting students to engage their prefrontal cortex is crucial for conceptual change to occur as a great deal of rational and reflective thinking is required to rework existing conceptual structures.

This is where mindfulness becomes even more interesting as a potential solution to the creationist | evolution conflict as studies using MRI scans have shown that there is a measurable reduction of the volume of amygdala after just one eight-week course of mindfulness practice. Moreover, as the volume of the amygdala shrinks, there is a simultaneous thickening of the prefrontal cortex which is associated with higher order brain functions such as awareness, concentration and decision-making (Taren et al. 2013). Other studies have shown that similar mindfulness training has led to increases in grey matter concentration specifically in those brain regions that are involved in learning and memory processes, emotion regulation, self-referential processing, and perspective taking (Holzel et al. 2011).

Moreover, researchers found that it was not only the *volume* of the amygdala and other brain regions that are changed through mindfulness but also the functional connectivity between the amygdala and other regions of the brain is altered through these practices. MRI scans show that the overall connection between the amygdala and the rest of brain becomes weaker, meaning that the amygdala has a much lower chance of hijacking the brain. Simultaneously mindfulness practices also strengthen the connections within other brain regions that are associated with attention and concentration (Taren et al. 2013). What this means according to neuroscientist Adrienne Taren is that "mindfulness practice increases one's ability to recruit higher order, pre-frontal cortex regions in order to down-regulate lower-order brain activity (Ireland 2014)", meaning that our more primitive responses to stress are then superseded by more intentional ones (Figure 21).

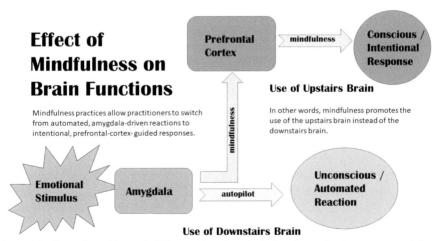

**Figure 21:** Illustration of how mindfulness affects how brain functions, allowing a movement from automatic, emotionally-driven reactions to conscious, intentional responses.

Color version at the end of the book

These types of neurological shifts can thus assist in lowering the emotional reactivity towards evolution, while also increasing the practitioners' overall sense of well-being. This increase in well-being and decrease in emotional reactivity is directly linked to the disempowerment of the amygdala. While the amygdala's ability to initiate the brain's rapid fight-or-flight response can save our lives when we are faced with imminent danger, it can also put us at risk for a broad spectrum of stress-related disorders if our bodies are exposed to repeated, excessive, or prolonged stress responses. In fact, the amygdala is now known to play a major role in mental health, with disorders such as depression and anxiety clearly associated with abnormally high amygdala function (Taren et al. 2013). These benefits could be even more pronounced for adolescents as the brain changes caused by mindfulness practices are directly correlated to the regions of the brain that are most closely linked to adolescents' difficulty in regulating negative emotional states (Blakemore 2014).

Let us put this into more simplified terms by turning to the idea of 'upstairs brain' and 'downstairs brain', where the downstairs brain represents all lower or more primitive functions such as innate reactions and impulses (e.g., fight-or-flight) associated with strong emotions (e.g., anger and fear) and the upstairs brain represents the more evolved portions of the brain responsible for higher mental processes, such as thinking, planning and rational processing of information (Figure 21).

Clearly it is advantageous for all individuals to be functioning using their upstairs brain when discussing complex topics such as evolution because it allows access their highest cognitive abilities, which are necessary for learning processes and conceptual change to occur. Using this metaphor of the upstairs brain and downstairs brain, we can see how that when a person is engaged in mindfulness practices, they are essentially using their upstairs brain to observe what is occurring downstairs. By engaging more often with the upstairs brain, individuals are better able to rationally process information and engage in higher mental processes. This is because, "In a state of mindfulness, thoughts and feelings are observed as events in the mind, without over-identifying with them and without reacting to them in an automatic, habitual pattern of reactivity. This dispassionate state of self-observation is thought to introduce a 'space' between one's perception and response (Bishop et al. 2004, p. 232)." This is paramount for evolution education as it means that mindfulness can allow members from each side to respond to situations reflectively as opposed to reflexively. In other words, by using the upstairs brain, both new atheists and creationists become more capable of participating in an effective discourse.

Moreover, it is possible to assist another person in the integration of the upstairs brain and the downstairs brain by responding and relating to that person in a manner that supports their ability to access higher brain functions such as reasoning and reflection. Here is worthwhile to look at Dr. Bruce Perry's work with traumatized children and the development of his three R's—regulate, relate and reason. According to Perry, regulation and relation are crucial foundational pillars for learning as a person first needs to be regulated in the sense that they feel physically and emotionally settled, so that they are capable of relating to the teacher or conversation partner (i.e., that they trust and feel connected to this other person). Only in this way will they be able to access their full mental capacity which is necessary for critical thinking, problem-solving, considering multiple solutions, etc. (Perry and Szalavitz

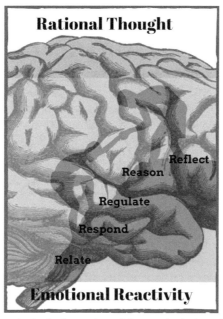

**Figure 22:** Teachers can assist in the integration of the upstairs brain and the downstairs brain by responding and relating to the student in order to help the student be able to access higher brain functions such as reasoning and reflection.

2011). Experts at the Department of Psychiatry Massachusetts General Hospital point out that this is not only true for traumatized children but for all children and even adolescents and adults (Think:Kids 2014).

This type of external help towards the integration of the upstairs and downstairs brain can occur both inside and outside of the classroom. It is possible to guide discussion participants to rational thought processes by first relating and responding to our conversation partner. This is essentially the opposite to the approach taken by new atheists, who continue to insult and demean creationists.

Now that it is clear how mindfulness can address and ameliorate the emotional reactivity surrounding the topic of human origins, we will look at how these practices can and have already been integrated into formal school environments.

## Mindfulness in Schools

Many of the mindfulness programs used for educational purposes have been loosely based on the MBSR program, which was developed by Jon Kabat Zinn (Table 6). The basic purpose or goal of a mindfulness practice is to recognize what is going on in the present moment (internally and externally) while taking on the role of the non-judgmental observer. In other words, a practitioner may observe their breath or label their thoughts and emotions as they come and go without judging them as good or bad or attempting to control them or change them. This type of awareness is described as "the ability to take in signals from the external environment as well as the thoughts and feelings…without getting stuck on any one stimulus to the detriment of others (Davidson and Begley 2012, p. 86)." The most interesting connection between mindfulness and science education is that mindfulness is analogous to intentional conceptual change in many ways as it reflects a voluntary state of mind that connects motivation, cognition and learning (Duit and Treagust 2003; Salomon and Globerson 1987).

It is thus not surprising that the interest in the mindfulness for educational purposes has increased exponentially. Nimrod Sheinman and Linor L. Hadar review the available research on mindfulness and offer further explanation for this increased attention in their contribution to the 2016 *Mindfulness and Education—research and practice*:

> *"In the last decade, interest in mindfulness with children (and adolescents) and mindfulness in schools has gone through a rapid expansion. The interest can be linked to the significant body of research documenting the clinical outcomes of mindfulness-based interventions with adults (Gotink et al. 2015), the contribution of mindfulness to healthy adults (Khoury et al. 2015), and the positive impact of mindfulness on children and youth (Weare 2012; Zoogman et al. 2015). The growing interest is supported by promising clinical outcomes of mindfulness with children and youth (Greenberg and Harris 2012), the feasibility of school-based mindfulness (Felver and Jennings 2016; Kuyken et al. 2013; Mendelson et al. 2010; Napoli et al. 2005; Weare 2012; Zenner et al. 2014), and the need for a more integrative and meaningful education intended to balance deficiencies and gaps in today's contemporary education (Davidson et al. 2012; Hart 2004; Zajonc 2016). In addition, the contribution of mindfulness-based practices to children seems to have potential in many nonpathogenic and education-relevant domains, such as attention training, emotional regulation, health promotion, stress management, social and emotional learning, compassion, resilience and wellness (see Schonert-Reichl and Roeser 2016 for updated reviews). As an experience, a life-skill and a resource for learning, mindfulness might be a unique asset to contemporary education (Sheinman and Hadar 2016, p. 78 & 79)".*

These findings relate directly back to what we have discussed in previous chapters as we have seen that the discussion of human origins is a complex, emotionally-charged subject. In the following sections we will look at how mindfulness-based applications can be used to create more positive learning environments in order to support the cognitive flexibility that can help inspire a move from conceptual conservatism to conceptual change.

While the majority of this chapter will be devoted to discussing how mindfulness can be used to address students' emotional states and mental flexibility, it is important to also take a look at the emotionality of teachers and how their emotional states also have a direct impact on students' cognition, motivation and behaviors (Sutton and Wheatley 2003), and therefore can either support or hinder the learning process. It has been found that the teaching of 'controversial topics' can cause negative emotional states in teachers and that "many teachers deal with highly stressful emotional situations in ways that compromise their ability to develop and sustain healthy relationships with their students, effectively manage their classrooms, and support student learning (Jennings and Greenberg 2011, p. 515)." Moreover, studies have shown that teaching 'controversial' subjects such as evolution requires a greater degree of emotional labor on the part of the teacher, which can lead to job dissatisfaction and even burnout (Swim and Fraser 2013). This can develop into a negative cycle that prevents teachers from effectively teaching and students from learning because teachers are more likely to experience negative affect when teaching evolution to students who openly reject evolution, and this negative affect will in turn effect that student's ability to process the information presented to them. This negative cycle can also increase the likelihood of a 'boomerang effect'.

While the boomerang effect is described within social psychology as the adoption of an opposing position in response to an attempt to persuade, here I use it to refer to an increase of belief perseverance in response to active teaching of data that contradict preexisting thought schemas. When the boomerang effect was first described in 1953, the authors pointed out that it was much more likely to occur when the information was associated with a negative source and when negative emotions were aroused (Hovland et al. 1953) and thus we can also imagine that belief perseverance is much more likely to occur in classroom settings where the teacher is perceived as a negative source or during discussions on evolution with an antagonizing conversation partner.

As we have discussed, the 'debate' over human origins occurs in two different realms for each side. While academics and secularists see the creationism | evolution conflict on the basis of rationality, data and facts, this conflict occurs in the realm of feelings and faith for the creationist. The importance of understanding evolution is paramount in the development of scientific literacy and we therefore need to support all members of society in their ability to understand this theory. It is therefore imperative that we create positive learning environments and space for discourse that (1) reduce the likelihood of a fight-or-flight response in students and (2) encourage cognitive flexibility and open-mindedness as the foundation for subsequent conceptual change. Mindfulness is not only theoretically helpful in achieving these goals, but the efficacy of mindfulness in education in changing learning environments through teacher training has already been tested and substantiated in a number of empirical studies.[19] In their article, Hen and Sharabi-Nov described the main findings from multiple studies:

- Kremenitzer and Miller (2008) found that teachers who can regulate their emotions and behavior are better able to manage their relationship with others and thereby facilitate positive classroom outcomes.

- Richardson and Shupe (2003) found that teachers with increased self-awareness had a more accurate understanding of how their emotional processes and behaviors are affected by their students, and how their emotions and behaviors, in turn, affect the students as well.

---

[19] For a detailed overview of how such a study works, please see the description of the Kemeny et al. study in the last chapter.

- According to Schutz et al. (2006) emotional competence plays a significant role in a teacher's ability to teach effectively.
- Chang (2009) found that emotional competence prevents burnout in teachers.
- According to Cohen (2006) teachers' emotional competence contribute to students' learning and performance.
- Gibbs (2003) proposed that teachers need to be able to regulate their internal world (emotions, thoughts and beliefs) in order to improve their teaching abilities.
- Sutton and Wheatley (2003) argue that lessons on the role of emotions in teaching should be an integral part of teacher training as teachers' emotions influence both their own and their students' cognition, motivation and behaviors in the classroom.
- Shapiro (2009) argues that teachers can develop a greater degree of solidarity and heightened sensitivity toward their colleagues and students through increased awareness of emotional experiences.

Based on the vast amount of research that has been amassed on the effectiveness of mindfulness, Jennings and Greenberg argue that "mindfulness-based interventions may be ideally suited to support the development of a mental set that is associated with effective classroom management (2011, p. 511)". This effect of mindfulness appears to be due to the fact that these practices allow us to reflect on our internal and external experiences and thus support our cognitive ability to regulate emotions, which in turn promotes an increased amount of flexibility in the way that we interpret and respond to stressful situations or topics (Zelazo and Cunningham 2007).

## Mindful Students and Conceptual Change

While positive learning environments and discourse can create conditions that reduce the likelihood of a fight-or-flight response or the boomerang effect, it does not necessarily elicit conceptual change. This is because the first step towards this reworking of conceptual structures is a person's willingness to engage with new material and to question pre-existing thought schemas. As stated earlier, Sinatra et al. has found that the processes of attitude change and conceptual modification are significantly impacted by student motivation and their emotional state and that the presentation of facts is not likely to be enough to cause students to relinquish their preconceptions in order to accept newly presented evidence (2014). As research has shown, there is a clear relationship between emotions and cognitive ability, indicating that positive emotional states are associated to an increase in cognitive functioning, while negative emotional states are associated with a narrowing of the cognitive functioning (Davidson and Begley 2012).

While one means of increasing positive emotional states can be achieved through the creation of prosocial classrooms, we must also recognize the fact that a negative internal emotional environment can also cause conceptual conservatism regardless of a teacher's emotional competencies or a friendly and welcoming learning environment. It is therefore imperative that we also address the ways in which we can support students in regulating their emotions, particularly when learning about evolution in the context of a high school class as we have seen that (1) the topic of human origins is inherently emotional (not intellectual) for creationists, (2) adolescents are prone to intense emotional states and have limited means of regulating these emotions, (3) intense negative emotions can prevent conceptual change. Thus, in order for conceptual change to occur, we do not only require positive learning environments where creationist students do not feel threatened, but we also need to provide these students with the cognitive tools needed to take the first steps towards conceptual change.

## Heuristics as Tools for Conscious Engagement in the Classroom

While many would like for the creationist to jump (quickly and directly) from denial to acceptance of evolution, creationist students often react negatively to the teaching of evolution and are resistant to conceptual change. As discussed, the creationist's negative reaction to evolution and their desire to reject evolution is understandable from an evolutionary standpoint and from a social and psychological perspective. Even though the existence of these negative emotional states is understandable, they are nevertheless undesirable as they are a hinder a person's ability to process information presented in the classroom and must therefore be addressed.

In this section we will look at specific mindful practices that can be taught to students to support their ability to learn, even when faced with emotionally difficult concepts. While these practices could be used for all types of emotionally salient subjects in schools, we will retain our focus on science education and evolution. In this context, mindfulness practices act as a tool for students to better recognize their own emotional states, thought processes, perceptions and judgements. This ability to become aware of one's own emotions, thoughts and judgements offers the critical space needed for students to regulate their emotions, thus gaining the ability to choose between an emotional reaction or a conscious response to a particular stimulus—such as a lesson on evolution.

A group at the City University of New York (CUNY), led by Kenneth Tobin, has already begun amassing data on the effectiveness of mindfulness in addressing and ameliorating emotions in the classroom. This group of researchers, in particular Kenneth Tobin, Malgorzata Powietrzynska, and Konstantinos Alexakos, set out to "develop interventions that would allow teachers (and students) to monitor their emotions, to counteract their negative impact and to maintain wellbeing (Powietrzynska 2015, p. 2)." This group of education specialists are convinced of the power of mindfulness in its ability to raise self-awareness and thus act as a powerful tool in shaping our emotional states (Powietrzynska et al. 2014). They recognized that one means of introducing mindfulness into the classroom is in the form of a heuristic, which is "a way of engaging in scientific search through methods and processes aimed at discovery; a way of self-inquiry and dialogue with others aimed at finding the underlying meanings of human experiences (Moustakas 1990, p. 15)" (Moustakas 1990; Powietrzynska et al. 2014).

The advantage of using heuristics is that they are flexible and adaptable and thus easy to apply to various contexts (Powietrzynska 2015). The efficacy of this approach is supported by studies conducted on reflexive sociology by Pierre Bourdieu and Loic J.D. Wacquant on reflexive inquiry, where reflexivity is understood as the process of being aware of aspects of oneself or environment that were formerly unaware (Bourdieu and Wacquant 1992). The CUNY mindfulness heuristic was designed to reflect seven facets of mindfulness as described in the table below (adapted from Powietrzynska's 2015 *Meaning of the Seven Facets of Mindfulness*, based on definitions offered by Ruth Baer, Erin Walsh, and Emily Lykins (2009)).

**Table 7:** Overview of the Seven Facets of Mindfulness from Powietrzynska based on definitions from Baer, Walsh and Lykins.

| Mindfulness Component | Description |
| --- | --- |
| Observing | Noticing, paying attention to internal and external stimuli such as physical sensations, emotions, smells, etc. |
| Describing | Verbal labeling of observed experiences, such as 'thought' or 'smelling'. |
| Acting with awareness | In contrast to 'automatic pilot' this describes actions that are done intentionally, actively and consciously. |
| Non-judging of inner experience | Refraining from evaluating the observed experiences; neutral stance, simple observer. |
| Non-reactivity to inner experience | Allowing thoughts and emotions to arise and dissipate without chasing them or getting caught up in them. |
| Curiosity | Interest in internal experiences, such as thoughts, emotions, etc. |
| De-centering | The ability to be aware of thoughts or feelings or experiences without identifying with them, i.e., understanding that we are not our thoughts. |

The development of these mindfulness characteristics, particularly 'acting with awareness' and 'non-reactivity to inner experience' are components that are particularly interesting for the teaching of evolution as they could act as the groundwork for students as they become aware of their own conceptual conservatism and begin to become aware of these belief systems without identifying with them. The resulting heuristic contained five characteristics for each of the seven facets of mindfulness. An excerpt of this initial heuristic can be seen below (Table 8) and a full version has been included in Appendix II.

As we can see, characteristic #4 (I don't criticize myself for having irrational or inappropriate emotions) corresponds to facet non-judging of inner experiences, while characteristic #6 (I have a hard time separating myself from my thoughts and feelings) corresponds to decentering. It is important to realize that the heuristic is written in a way that does not prescribe what is 'expected' or 'desired' but instead just presents possible states, allowing the user to develop awareness of their current state of engagement and behavior, while also recognizing the possibility of other possible actions.

The statements in the heuristic can then be coupled with a rating or frequency scale and respondents are then asked to choose the rating or frequency that best

**Table 8:** Excerpt of the first draft of the CUNY mindfulness heuristic with seven facets of mindfulness.

| Characteristics in the First Iteration of the Mindfulness Heuristic |
|---|
| 4    I don't criticize myself for having irrational or inappropriate emotions. |
| 5    I perceive my feelings and emotions without having to react to them. |
| 6    I have a hard time separating myself from my thoughts and feelings. |
| 7    I am not curious to see what my mind is up to from moment to moment. |
| 8    It is hard for me to put my beliefs, opinions, and expectations into words. |
| 11    When I have distressing thoughts or images, they tend to consume me. |
| 14    I am not curious about my thoughts and feelings as they occur. |
| 16    I make judgments about whether my thoughts are good or bad. |
| 17    In difficult situations, I can pause without immediately reacting. |
| 23    I approach my experiences by trying to accept them, no matter whether they are pleasant or unpleasant. |
| 24    I am curious about my reactions to things. |
| 27    I am curious about what I might learn about myself by just taking notice of what my attention gets drawn to. |
| 29    I tend to react strongly to distressing thoughts and/or images. |
| 30    When I have distressing thoughts or images, I judge myself as good or bad, depending what the thought/image is about. |
| 33    I am aware of my thoughts and feelings without over-identifying with them. |
| 35    When I have distressing thoughts or images, I just notice them and let them. |

reflects their current behavior or mindset. In theory, as each respondent reads the outlined characteristics, they are exposed to various features of mindfulness and emotional styles. In principle, it does not matter whether their ranking is 'correct'. In other words, if an independent observer would say that the respondent displayed a particular behavior *seldomly*, yet the respondent marks *frequently*, it does not change the validity of the heuristic as the heuristic is designed to be a reflexive tool and not a psychometric. As the CUNY group states, "Heuristics are meant to afford changes in the way a person enacts social life… . As a person reads a characteristic, in addition to becoming aware of a particular feature of mindfulness, he or she makes a determination as to the extent this characteristic is salient to his/her ontology. During this reflective process, certain characteristics in the heuristic (more than others) may resonate with the respondent. In the process, he or she may consider adopting practices that would be aligned more closely with those characteristics. In addition to the availability of the rating scale to indicate the frequency of occurrence of certain characteristics in one's social life, we provide space for additional reflections over each characteristic (Powietrzynska et al. 2014, p. 71)." An example of the revised heuristic with rating system can be found in Appendix II.

For our purposes, we can see that a student with belief perseverance could read characteristic #6 and begin to realize (possibly for the first time) that they do closely identify with their thoughts and emotions and simultaneously recognize that it is not necessary to do so. Likewise, both secularists and creationists could read characteristic #17 (*In difficult situations, I can pause without immediately reacting.*) and reflect upon the way that they have reacted to fellow students and/ or the teacher during a heated discussion about human origins or the validity of evolution. Theoretically, simply recognizing one's own emotional reactivity begins an integration of the upstairs and downstairs brain as this type of recognition requires an activation of higher cognitive powers to observe and analyze the 'behavior' of the downstairs brain.

In addition to basic mindfulness, the CUNY research group also decided to incorporate the work of Richard J. Davidson and Sharon Begley on *emotional styles* and Paul Ekman's work on *emotional profiles*. Davidson's work on emotional styles is based on his research in affective neuroscience and his observations regarding the different ways that individuals react to similar life events. As Davidson states, "Emotional Style influences the likelihood of feeling particular emotional states, traits, and moods. Because Emotional Styles are much closer to underlying brain systems that emotional states or traits, they can be considered the atoms of our emotional lives— their fundamental building blocks (Davidson and Begley 2012, p. loc. 82)" . Davidson proposes six main emotional styles dimensions: resilience, outlook, social intuition, self-awareness, sensitivity to context and attention (Table 9).

The CUNY group saw immense potential in the application of Davidson's work for educational purposes because his research demonstrated that these particular emotional styles could be altered through contemplative practices, such as mindfulness, due to the brain's neuroplastic character. Davidson showed for example that resilience could be strengthened through mindfulness due to the increase in connection between the left prefrontal cortex and the amygdala. In addition to resilience, the CUNY group recognized that changes to a person's outlook and attention would have a vast effect on learning, particularly with regard to a person's ability to remain focused or maintain positive emotion.

The benefits of addressing these different emotional styles also has very direct implications for teaching evolution to creationist audiences, particularly with regard to the development of resilience, as we have seen that disrupting literalist interpretations of Genesis may lead to an unintended destabilization of meaning systems and terror management strategies. The destabilization of these psychological frameworks can naturally lead to states of anxiety and depression and thus it is desirable and ethical to want to provide these students with increased resilience to support them in overcoming these initial emotional setbacks. We have already discussed the potential parallels between the grief process and the pathway from evolution-denial to evolution-acceptance and the development of these emotional skills could prevent students from getting 'stuck' in the depression stage (Figure 15).

Table 9:  Six Emotional Styles According to Davidson and Begley.

| Emotional Style Dimensions (Davidson and Begley 2012) | |
|---|---|
| Resilience | How quickly you recover from adversity. |
| Outlook | How long you are able to sustain positive emotion. |
| Social Intuition | How adept you are at picking up social signals from the people around you. |
| Self-Awareness | How well you perceive bodily feelings that reflect emotions. |
| Sensitivity to Context | How good you are at regulating your emotional responses to take into account of the context you find yourself in. |
| Attention | How sharp and clear your focus is. |

The CUNY group translated Davidson's emotional style dimensions into heuristic characteristics such as (Resilience) *I quickly recover when things go wrong for me*; (Outlook) *I maintain a positive outlook on life*; (Social Intuition) *I can tell when something is bothering another person just by looking at him/her* (Self-Awareness); *When I am emotional, I notice changes in my heartbeat* (Sensitivity to Context); *The extent to which I show my emotions depends on where I am* (Attention); *If I decide to focus my at a particular task, I can keep it there.*

They then decided, however, that a mindfulness heuristic would not be truly complete without another component of traditional mindfulness practice known as *lovingkindness*. Lovingkindness or so-called Metta meditations are common within certain Buddhist traditions[20] and are also used as part of formal therapy programs, such as mindfulness-based cognitive therapy (MBCT) and dialectical behavior therapy (DBT). These practices are also becoming increasingly popular for non-spiritual and non-pathogenic purposes such as general well-being and stress-reduction. Regardless of the place where these practices are exercised, the ultimate goal of such practices is to induce an open and compassionate stance towards all living beings, including oneself. The importance and ramifications of this type of compassion practice are enormous, not only in terms of science education but for society at large—particularly in the context of discussions of 'controversial' topics (evolution, climate change, etc.) as well as socially-sensitive topics (sexual orientation, abortion, etc.).

With the incorporation of lovingkindness and emotional styles, the CUNY group thus defined nine dimensions of mindful action or activity: (1) Being aware of surroundings, emotions and what you are doing, (2) Maintaining focus (3) Being kind (4) Acting with compassion (5) Recovering from adversity (6) Maintaining a positive outlook (7) Being socially intuitive (8) Adapting actions to context (9) Separating emotions from other actions. All nine of these dimensions are particularly relevant to teaching and discussing evolution as it addresses the necessary components for supporting conceptual change in the form of positive learning environments, and willingness to engage with new material, while also providing students will the tools to deal with the potential negative emotional byproducts of restricting previous conceptual structures.

Lastly, the overall importance of social dynamics was reflected in the CUNY group's decision to incorporate heuristic components that specifically addressed the

---

[20]   In formal meditation, a lovingkindness practices often take place in the form of a chant or silent meditation where certain statements are repeated mentally or out loud in an attempt evoke a sense of lovingkindness or compassion or general friendliness towards oneself and others. In typical practice situations, statements are first directed at oneself, then towards a so-called 'benefactor' (someone to whom you have a nice and easy relationship), then to a neutral person and finally to a difficult person (or someone that the practitioner experiences as difficult). These repeated statements are often a simple well-wishing of a person to be free from harm and pain, etc. An example of such phrases include: *May I be safe. May I be happy. May I be healthy. May I be free of suffering.* Likewise, statements aimed at others have the same sort of compassionate-wishing and can include almost identical phrasing, such as: *May you be safe. May you be at ease. May you be free from suffering.* While the repetition of these statements can feel mechanical or unnatural in the beginning, overtime practitioners theoretically develop a larger degree of compassion and empathy making it easier for these statements to feel authentic and to bring them joy.

classroom environment, such as: *I am aware of emotional climate and my role in it*. Moreover, to assist in the development of emotional awareness, characteristics were developed to help individuals become aware of one's own emotions through the observation of physiological processes, such as pulse rate, body posture and body temperature, while other characteristics were developed to bring attention to the expression of emotions in voice, face, and body movements to support the development of friendlier relationships in the classroom. All of these new ideas (emotional styles, loving kindness, emotional regulation, social interactions) were then incorporated into a new heuristic. An excerpt of this heuristic can be found below (Table 10), a full version is available in Appendix II.

We can see from characteristics #7, 23, 31, for example, that the CUNY group also integrated components related to the link between physiological processes and emotions as well as the ability to regulate emotions through such processes, such as breathing exercises. This link between physiological processes is important as it has been found that "emotional thoughts, either conscious or nonconscious, can alter the state of the body in characteristic ways, such as by tensing or relaxing the skeletal muscles or by changing the heart rate. In turn, the bodily sensations of these changes, either actual or simulated, contribute either consciously or nonconsciously to feelings, which can then influence thought (Immordino-Yang and Damasio 2007, p. 8)."

The goals of the heuristic appear to have been met as participants stated that "the heuristic made them think and internalize their feelings; made them stop and think more and be more reflective than they usually were; made them think of things they never thought about; made them think about themselves. Thus, it was evident that the heuristic successfully mediated reflexivity since it actually worked as an enhancer of self-awareness (Powietrzynska 2015, p. 4)." As Tobin points out, this type of reflexivity is crucially important because much of social life occurs without conscious awareness and such practices allow actors to better identify the way in which they are affected by their environment and how they affect other players in

**Table 10:** Excerpt of mindfulness heuristic from CUNY group that addressed emotional styles, classroom dynamics and the ability to regulate emotions through physiological processes.

| Characteristics in the Mindfulness in Education Heuristic (Powietrzynska 2015) |
|---|
| During this class: |
| 3    I identify distracting thoughts but let them go (without them influencing future action). |
| 5    I recover quickly when I am unsuccessful. |
| 7    I am aware of the relationship between my emotions and breathing pattern. |
| 9    I maintain a positive outlook. |
| 11   I can tell when something is bothering other students. |
| 14   I can focus my attention on learning. |
| 16   I feel compassion for others when they are unsuccessful. |
| 17   When I produce strong emotions, I easily let them go. |
| 20   I am aware of my emotions being expressed in my voice. |
| 21   I recognize others' emotions by looking at their faces. |
| 23   My emotions are evident from the way I position and move my body. |
| 24   The way I position and move my body changes my emotions. |
| 26   I am aware of emotional climate and my role in it. |
| 28   Classroom interactions are characterized by winners and losers. |
| 31   I use breathing to manage my emotions. |

this environment and also offers these actors a means by which they can rationally change aspects of their behavior to benefit the collective group (2012). These points are directly applicable to the creationism | evolution challenge as we have seen that belief perseverance on the part of the creationist is often driven by intensely emotional, yet subconscious factors, such as a desire to belong to a particular community or a means of reducing general anxiety. The cognitive flexibility that is gained through such practices are not only beneficial for more caring and supportive learning environments but also generally support a state of open-mindedness and thus lay the groundwork for more effective learning and potential conceptual change.

Another means of applying heuristics or contemplative practices to ameliorate tensions around 'controversial' topics is exercises focused on promoting mindful speech and mindful listening. Here the word mindful could also be replaced by respectful or kind. As Alexakos and colleagues state, "Accepting difference, even when opposed to ours and showing compassion and at the same time not being hurtful, is not easy to achieve in a class. Mindfully speaking and mindfully listening, being aware of each other's emotions, acting together, exploring each other's views without having to agree and without being threatened are some practices that encourage social entrainment and the building of solidarity (Alexakos et al. 2016, p. 763)." A full version of Tobin, Alexakos & Powietrzynska's proposed mindful listening and speech heuristics, as well as Alexakos et al.'s heuristic specifically designed for discussion of 'thorny' issues can be found in Appendix II.

While the CUNY group decided to promote mindful listening and speech in the form of a heuristic, these same characteristics could also be used as a set of guidelines for ethical classroom interaction and general discourse as the implications and benefits of these guidelines are clear with regard to discussing emotionally-salient topics, such as evolution. These guidelines are also useful in general, though, due to their ability to ameliorate the tension that arises in discussions due to differing political stances, worldviews, opinions, etc.

## Summary

Mindfulness is in many ways analogous to conceptual change as it reflects a voluntary state of mind that connects motivation, cognition and learning. The most exciting data that have come from research on contemplative practices, such as mindfulness, is that these practices can actually be used to intervene in our own conscious evolution. As neuropsychologist Rick Hanson pointed out, we are able to reconstruct and reorganize our brain structures through self-directed neuroplasticity.

With a specific look at the creationism | evolution conflict, mindfulness can support individuals on both sides of the conflict in their ability to recognize when their perceptions and reactions are being driven by primitive maladaptive reactions. Simultaneously, such practices offer these individuals the ability to choose to engage with their higher consciousness so that their responses are guided by rational thought rather than being driven by more primitive limbic reactions.

The limbic system is particularly active when creationists are confronted with discussion on evolution. While secularists often see the topic of human origins as an intellectual question, it is a highly emotional topic for the creationists. Despite the negative emotions that arise in connection with evolution, it is imperative that all members of society understand the principles of evolution as it is central to the development of general scientific literacy. To support all individuals in their ability to understand and accept evolution it is imperative that positive learning environments and space for discourse are created that (1) reduce the likelihood of a fight-or-flight response in students and (2) encourage cognitive flexibility and open-mindedness as the foundation for subsequent conceptual change. Mindfulness is not only theoretically helpful in achieving these goals, but the efficacy of mindfulness in education has already been tested and substantiated in a number of empirical studies.

In general, the development of self-awareness is crucially important since much of social life occurs without conscious awareness (Tobin 2012). A general increase in self-awareness would greatly support effective discourse on evolution as it allows both sides to better identify how they are affected by their environment, how they affect others around them and how they can rationally change aspects of their behavior to benefit the collective group and support the efficacy of the dialogue. Moreover, the general non-judgmental stance which is promoted in mindfulness practices creates a space between a person and their beliefs that allows them to better recognize their own conceptual conservatism and how to identify less closely with fundamentalist belief systems.

At the same time, if educators and science communicators are able to take on this peaceful and non-judgmental stance, it supports them in their ability to develop a learning environment where creationists feel safe enough to listen to the data presented to them and to begin to question their previous conceptual structures. Mindfulness can therefore be seen as a key piece of the science literacy puzzle and in the next chapter we will look at ways that these practices can be used to specifically support effective discourse on human origins. More information on how mindfulness practices can be implemented in the classroom can also be found in Appendix IV.

# Chapter 7
# Education (R)evolution

As I stated in the preface, this entire book can be understood as a very extended answer to a question that I received during my doctoral defense. My doctoral research had focused on the creationism movement in the United States and its effect on the public-school education system. During my defense I highlighted the creationists' willingness to go to bat in order to get their worldviews and ideologies introduced into the science classroom. Towards the end of my defense one of the members of my panel asked, "So what should a teacher do when they are confronted by a creationist student?" I had studied the creationist phenomenon for over four years by that point, but I was still unprepared for the question. Luckily, I was able to answer the question to his satisfaction by pointing out that studies had shown that the use of cladograms and examples of megaevolution can be very effective in explaining evolution to students (Padian 2010). The panel was happy with my answer, but I was not. I had sat in classrooms with creationists and I knew that a better explanation of evolution alone would not suffice in addressing the perpetuated misconceptions that these students had brought with them to the classroom.

As discussed, correcting a person's misconceptions about the natural world requires a person to reevaluate, restructure or relinquish their previously held conceptual structures in favor of the scientific explanation presented to them. This process is known as conceptual change and it is often hindered by conceptual conservatism or belief perseverance. The only way to overcome this conceptual conservatism, is to recognize and address the involvement of emotions in a person's choice to uphold certain misconceptions. Recognizing the role that emotions play in science education allows us to better understand why better illustrations and better examples simply do not suffice as solutions. Any viable solution will need to address the intense emotional states that arise when learning about evolution because ameliorating negative emotional states is the first crucial step in opening the doors to conceptual change.

While the conflict over human origins may be purely intellectual for many scientists and many educators, we have seen that the issue runs much deeper for the creationists. We have seen that it is therefore naïve to believe that the creationist can jump from denial to acceptance without any emotional response. As I mentioned earlier in this book, the move from denial to acceptance can be equated with the process of grief (Figure 15). We have seen that religion in general, and fundamentalist beliefs to an even greater degree, act as a framework which provides meaning and direction for many individuals. When these individuals are then confronted by data that cause them to question their belief system (and ultimately their meaning system), it destabilizes their understanding of themselves and the world.

This is why we cannot expect for creationists to move from denying evolution to accepting evolution simply through an influx of data. We also cannot expect this change to occur over night. We need to realize that effective educational intervention is the equivalent to planting a seed that may require a great deal of time and special circumstances to take root. While we cannot force this movement from denial to acceptance to occur within a particular semester or conversation, we can provide an environment where the seed is much more likely to take root. In other words, positive learning environments, mindful practices that increase cognitive flexibility and a person's ability to reflect on their thoughts and emotions function like the fertilizer, water and sunlight that encourage the seeds of scientific literacy to take root and flourish.

In this way it is true that teaching megaevolution and the use of better diagrams can support the understanding and acceptance of evolution, but these types of teaching methods and tools are only effective if there is a willingness on the student's part to listen with the intent of learning. In the presence of conceptual conservatism all teaching methods are rendered useless unless cognitive flexibility is fostered in the students. In the rest of the chapter we will look at specific tools that can be used to address negative emotional states and conceptual conservatism in order to develop a greater degree of self-awareness among students and teachers, which thereby fosters not only cognitive flexibility but also pro-social learning environments. Once we have addressed the emotional nature of this issue and discussed specific exercises that can be used to alleviate and circumvent intense negative emotional states, we will also look at two classroom exercises that have been designed for students to come to a better understanding of evolution while promoting curiosity, scientific thinking and cooperation. While the focus here is on 'classrooms' these same ideas and techniques can be easily adapted and applied to all discussions on evolution that occur outside a formal educational environment.

## Mindfully Addressing Misconceptions and Increasing Science Receptivity

As we have seen, creationist thought is embedded our country's culture, history and society. It is also entrenched in many individuals' sense of self and the means by which they navigate their emotional lives. The deep roots of creationism both at a societal and an individual level is what makes it so hard to find a sustainable solution to evolution denialism.

As we began to discuss in the past chapter, cognitive tools, such as mindfulness, can be used to drive an intentional change in our mental modes, allowing us to essentially become the driver of our own conscious evolution. Here we will explore how we can take advantage of the brain's plasticity to develop a higher level of self-awareness that allows us to become more cognizant of our own thought processes and emotional reactivity to those thoughts. Essentially, by taking on the role of the 'observer', individuals, i.e., students, teachers, creationists, new atheists, etc., are moving from a 'downstairs-brain emotional reflexive mode' to an 'upstairs-brain reflective mode' as they attain greater cognitive flexibility and control over their reactions.

For example, creationists may become cognizant of their emotional reactivity to evolution and recognize the connection between this emotional reactivity and the certain 'belief statements' such as 'evolution is amoral' or 'doubting Genesis is the same as doubting God'. The ability to recognize the presence of these 'belief statements' and the subsequent emotional reactivity that these statements elicit creates space between the 'observer' and these statements thereby opening the door to intentional, rational behavioral choice. In other words, by recognizing these belief statements as such, creationists then have enough 'space' between themselves and these statements to decide whether or not these statements are actually true, whereby they then gain the freedom to choose their response to these statements. When they are not aware of option of 'believing' these statements, they are basically trapped 'downstairs' with little access to their higher cognitive powers. In essence, they are running on autopilot where as soon as a lesson on evolution is presented, they 'automatically' shut-down or become defensive. Simply by recognizing these belief statements, one makes the critical move 'upstairs' in order to take on this role of the 'observer' and this move allows the cognitive freedom to 'choose' to shut-down or to listen to the data presented to them.

Likewise, new atheists may recognize their tendency to judge creationists as ignorant and subsequently also achieve enough mental space from their judgmental thoughts to actively and consciously choose whether or not insulting the creationist is the most effective means of communicating about the validity of evolution. The presence and distracting nature of these thoughts is so ubiquitous that we are often unaware of their presence and their ability to hijack our mind and taint our perception of the world, as Gil Fronsdal points out "The mind can be so 'distracted by distractions it does not even know it is distracted' (2008, p. 99)."

This shift from habitual, automatic (unconscious) reactions to intentional, conscious responses is a core idea of cognitive therapies such as DBT (Table 6). During these types of therapy sessions, the focus is clearly on a very different set of 'belief statements' and habits such as drug dependency, but the basic principle is still very useful for non-pathological applications. Patients in these types of programs receive extensive 'skills' training based on mindfulness practices in order to (1) recognize unhealthy thought processes and behavioral habits and (2) recognize the choice to follow alternative thought processes and habits.

This recognition of choice is a crucial step, but the process of changing habitual thought and behavior is a neurological uphill battle as these habits are hard-wired in the brain and act like high-speed highway where driving is quick, easy and comfortable. The pursuit of new behavioral trends on the other hand, can be equated with exiting this highway and attempting to drive along a forest path at high-speed, which can result in a bumpy ride to say the least. But due to the plastic nature of our brain, the practicing of these new behaviors can cause a rewiring of the brain in which old pathways are degraded in favor of strengthening new pathways. Meaning that over time, the ride along the highway becomes slower and the exit onto the forest path becomes smoother and the more often one chooses to exit the highway, the smoother the new path through the forest becomes. At a neurological level, what ends up happening is that the shift from old habits to new habits becomes easier as the potency of the old habits is lost through a lack of reinforcement. Overtime the forest path becomes a road and the old highway succumbs to potholes.

Based on these same principles, we can expect that while mindfulness practices may allow a creationist student to recognize the presence of belief statements about evolution that are grounded in fundamentalist thinking and may support their ability to identify the negative effects that these statements have on their learning potential, the decision to abstain from following these thoughts is still difficult, especially in the beginning. Yet, every time that a person does not *immediately* shut-down when they hear the world 'evolution', they take one step away from conceptual conservatism and one step towards scientific literacy. The good news here is that the brain is primed for this type of change during adolescence as the brain is already set to prune and myelinate.

In the following sections, I will offer an overview of specific tools that are designed to help individuals become aware of how their perspectives, judgements, thoughts and views affect them emotionally and hinder rational thought. The purpose of these tools is also to create safer learning and communication environments, where both sides can actively work on reducing their tendency to 'other' in order to hear and accept each other—regardless of their disparaging worldviews.

The set of tools consists of four main parts: (1) classroom guidelines that act as a basis for respectful dialogues, (2) specific mindfulness practices that allow a person to become more aware of their thoughts, emotions, opinions and relationships during the discussion of human origins, (3) specific stress-relief tactics such as breathing exercises that can be used to support mindful awareness and subsequently respectful dialogues and lastly, (4) specific classroom lessons designed to shift gears from fear-aversion to goal-seeking behavior. We will examine the first two components in this section and then move on to examine breathing exercises and gear-shifting lessons later in this chapter.

## Classroom Guidelines

As we have seen, the only way for conceptual change to occur is for a person to be able to engage with new material and to be willing to reevaluate existing conceptual structures. Aggressive speech or overt attempts to persuade these students will most likely lead to cognitive shut-down and in the worst case a boomerang effect. The necessity of positive learning and communication environments is essential and must be established as the groundwork for future conceptual change.

For this purpose, it is sensible to create a general set of rules for respectful classroom dialogue. Here we will look at specific examples for educational environments and then look at more general means of promoting effective dialogues at a societal level. These classroom guidelines could be established at any point in the school year as pro-social classroom environments support all learning. The guidelines should not only address the manner of communication between students but also between the teacher and students. When teaching about evolution, it is important that the guidelines explicitly point out what can and cannot be discussed in a science classroom from a legal perspective (Table 11).

Within these guidelines we can see that there is clear language to remind students that a pro-social classroom cannot be equated with an 'anything goes' situation. An empathetic stance towards creationist students does not mean that we are 'flexible' with regard to science standards, nor are we willing to accept intelligent design as an 'alternative' theory. Within the science classroom—science needs to remain science. At the same time, in our interpersonal communications—humans need to remain human.

In other words, while we can show the creationist *student* a great deal of compassion and understanding, this degree of flexibility and openness does not apply to the creationist movement. Here we must be very clear that neither creationism, nor intelligent design belongs in the science classroom. What this means from a practical point is that we strictly prohibit the introduction of creationism into the classroom while simultaneously refraining from demonizing the creationist student. In other words, we have a no tolerance policy on attempts to undermine science education, but in response to our students, we place their humanity before their ideology. In this way, we show the creationist students compassion, while still maintaining educational integrity and ensuring that due respect is given to the teacher.

While these general classroom guidelines can be very useful for students who already have a certain degree of empathy, it is also possible to introduce additional, specific ten-point guidelines that promote kind and respectful communication in the form of mindful speech and listening (Table 12), which were adapted from the materials developed by the CUNY group (Appendix II). In comparison to the proposed classroom guidelines, the universal nature of these guidelines makes them particularly beneficial for all types of discussions on evolution due to their ability to reduce tensions and division. Clearly, we cannot walk around with a copy of these guidelines to hand out to creationists we encounter and request them to abide by the same guidelines, but more often than not, it suffices if one person abstains from negative speech, encouraging the other to follow suit.

**Table 11:** Classroom guidelines for discussing evolution or other 'controversial' or emotionally difficult topics in the classroom.

| **Classroom Guidelines for Respectful Communication and Interactions** | |
| --- | --- |
| *During lessons on evolution or emotionally-salient topics in the classroom* | |
| 1 | I realize that my teacher's job is to teach us the content of the school curriculum and is required by constitution of the United States to uphold the separation of church and state which prevents teachers from teaching religious content in science classrooms. |
| 2 | I respect my teacher and my fellow classmates and acknowledge and respect the fact that they may hold very different views of the world than my own. |
| 3 | My focus in the classroom is on learning, not only from the teacher but also from my fellow classmates. |
| 4 | I attempt to remain open-minded, attentive, kind and engaged to support my own education and that of my classmates. |
| 5 | Even when disagreements arise, I will attempt to promote harmony instead of discord. |

**Table 12:** Guidelines for respectful speech to support pro-social classroom interactions and general discussions on evolution.

| Guidelines to Increasing Understanding through Mindful Speech |
|:---|
| *When I participate in classroom discourses* |
| *Mindful/Respectful/Kind listening:* |
| 1   I show my respect for the speaker, while trying to understand and remain open-minded. |
| 2   I attempt to cultivate compassion and empathy towards the speaker, even when I do not agree with them. |
| 3   I attempt to listen and learn, even when the opinions presented are not in line with my own. |
| 4   I abstain from judging the speaker, realizing that all humans have different backgrounds that feed into their own perspectives. |
| *Mindful/Respectful/Kind speech:* |
| 5   Before speaking, I pause to make sure the previous speaker has finished. |
| 6   I look for signs that others want to speak and end my turn to avoid monopolizing the discussion. |
| 7   As I speak, I monitor my emotions and attempt to regulate strong negative emotional reactions as not to let them affect my speech. |
| 8   I do not increase the loudness of my voice to overpower other classroom participants, respecting that every student has a right to participate in the conversation |
| 9   As I speak, I attempt to convey a specific message while consciously refraining from causing harm to the listeners |
| 10  Above all I engage with my fellow classmates with an open heart and an open mind and my speech is free of hurtful or divisive rhetoric. |

Both sets of these guidelines can act as the basis of a pro-social classroom environments and as the basis of respectful dialogues outside of the classroom. This feeling of safety and respect can prevent the triggering a boomerang effect or a fight-or-flight response. This sense of safety within the classroom can be further supported by the establishment of a general mindset guided by (radical) acceptance and validation. Both radical acceptance and validation are components of formal DBT therapy as part of the skills-set designed to lower stress and improve interpersonal relations. For our purposes, validation and acceptance can act as the cornerstone for emotionally-intelligent teaching and conversation models.

Before continuing, one point should be made very clear: acceptance and validation should not be equated with condoning or agreeing. When I speak of acceptance and validation, I am referring to the skills that have been implemented within the framework of MBCT and DBT (Table 6). I do not mean that we should *accept* the teaching of creationist beliefs in the classroom, nor do I mean that we should *validate* these beliefs in the sense of declaring them legitimate. In other words, we do not pretend to find the idea of a 6,000-year-old Earth reasonable, nor should this be misunderstood as encouraging the teaching of 'both sides'. Instead, acceptance here refers to the ability to 'recognize' and 'accept' the true nature of a situation, while validation is focused on 'honoring' the humanity of an individual and their personal experience of a situation.

Acceptance in this sense is the ability to recognize and accept the true nature of reality, while also recognizing what action is possible within the current situation. Carl Rogers, founder of MBCT, emphasized that acceptance is a precondition of change, i.e., that a change in thought patterns or behavior is dependent upon a profound acceptance of the experience. The idea of acceptance or radical acceptance is also central to other mindfulness-based therapies. According to the creator of DBT, Marsha Linehan, there are four options when you have a problem: (1) solve the problem, (2) change your perception, (3) radical acceptance or (4) be miserable (Linehan 1993). Here 'radical acceptance' is defined as the willingness to take life at it is, without aversion or judgement. In other words, one let's go of ideas of 'how things *should* be' and simply sees the reality as it *is* in the present moment. You accept the fact that you do not have the means or opportunity to change the situation *at present*. The core idea behind radical acceptance is a form of reducing negative emotions in difficult situations and to encourage effective action. For example, imagine a patient who is an alcoholic, the only means of living a healthy life as a recovering alcoholic is abstinence. This decision to completely abstain from alcohol requires the patient to accept the fact that they are an addict. Here we see acceptance precedes effective action. At the same time, radical acceptance helps us deal with difficult situations where there is no easy or immediate solution available.

From an educational standpoint, this means that American teachers are able to recognize and 'accept' the current reality in America, i.e., there are many Americans who avidly oppose evolution. This does not mean that we 'approve' of this opposition but that we are able to rationally recognize that this is the current state of affairs. In accordance with this reality, educators can reasonably expect that there will be creationist students who also vehemently reject evolution. This is our current reality. The reason why acceptance is so important is because accepting the reality of a situation is the first critical step towards recognizing what can or should then be done within a particular set of situational conditions. We have to *accept* reality, so we can *deal* with reality.

What this means is that we cannot simply deny the existence of creationism, nor can we simply 'wish' the creationist phenomenon away. We have to accept the presence of creationism so that we can develop effective means of becoming proactive and effective within the given situation. Radical acceptance does not change a situation, but it offers us enough emotional space to see what action is possible within the difficult situation. For this reason, I propose that 'radical acceptance' is very useful tactic when teaching evolution, particularly in the United States.

Again, it is important to remember that accepting this fact does not mean that we are saying that this situation is good or acceptable. Accepting in this sense is not to be understood as a type of judgement but is perhaps better understood as an attitude. It is simply accepting reality as it is with the purpose of gaining more cognitive flexibility and decreasing unnecessary stress. This type of acceptance offers a multitude of positive attributes for the teacher and the students. Radical acceptance allows the teacher to avoid additional painful emotions that may lead to decreased well-being; it has the potential to decrease classroom tensions and can prevent general job dissatisfaction. Most importantly, though, in terms of scientific literacy, radical acceptance allows us to recognize teaching opportunities that might have otherwise been overseen. It allows the teacher to ask herself—ok what is possible? What is the most useful way to teach evolution when I have students who are resisting everything I say?

Many may argue that the better option here is to 'solve the problem' as Nobel laureate Steven Weinberg stated, "Anything that we scientists can do to weaken the hold of religion should be done and may in the end be our greatest contribution to civilization (1984)." Steven Weinberg is not alone in his desire, as others such as

Richard Dawkins have made similar proclamations. Yet, these kinds of proclamations are akin to telling an addict that the 'solution' to the problem is simply not be addicted anymore. While philosophers may have the luxury of hypothesizing about a world without religion or a world without drug addiction, there is no such room for these types of daydreams in an educational setting. If teachers began to make a list of all of the different types of behaviors and types of students that they would prefer not to have in their classroom, the list would become very long. It is for that reason, that divisive, aggressive speech is not only useless but counterproductive. Whether we are in agreement with them or not, we will have to accept the fact that many individuals hold creationists viewpoints and are generally hostile towards science. We cannot wish these individuals away. Nor can we hold our breath waiting for this wave of fundamentalist activity simply to blow over.

As we saw in the chapter on the history of creationism and legal turmoil, this battle has been raging for over one hundred years. While I do believe that this fundamentalist trend can be dismantled, I do not hold any hopes that this will occur in the immediate future and possibly not even within this century. This means that we will have to accept this reality and better equip teachers at dealing with those who challenge them so that we can develop better methods to engage with those who continue to adamantly doubt the theory of evolution.

Moving on to the idea of validation, we find another means of reducing emotional stress while also deepening trust amongst individuals regardless of whether they have opposing worldviews or contrary religious beliefs. In day-to-day-life the word 'validation' is most often associated with the verb 'validate' in the sense of proving the accuracy of something or declaring that something is legally acceptable. To be very clear, I am in no way proposing that we should declare that creationist claims are valid in a legal, educational or scientific sense. Validation in this psychosocial context means affirming that a person, their ideas and their feelings are valuable and worthwhile. Like radical acceptance, validation is a skill that is often taught to DBT patients in order to help them regulate their interpersonal relationships by verbalizing that they value the other person as a human being. Validation at its core involves recognizing another person's feelings and not being dismissive of those feelings. It can also be used to validate one's own emotional experience of a situation.

Validation can be very useful tool with communicating about human origins because it is often very difficult for either party to understand the emotional state of the other. On the one side, the creationists fear that their children will be taught evolution and lose their faith. On the other side, we have secularists who are very concerned about having religious ideologies smuggled into the science classroom because they believe that it reduces a student's ability to discern the difference between scientific fact and faith and can therefore have deleterious effects on scientific literacy. In such a complex situation, it is easy to be dismissive of the other person's emotions as the opposing side develops into a mass of unreal others.[21]

---

[21]    The term 'unreal other' was coined by clinical psychologist Tara Brach as a description of the attitude that we develop towards those who do not belong to our 'tribe' or who we identify as being very different from ourselves. The more different a person or their circumstances are to our own, the more unreal they become.

So, when I speak of validation, I do not mean that we should legitimize ideas such as intelligent design, but instead that we should show that we recognize and value the creationist as a human being with emotions. While their *ideas* about evolution are misguided and often straight-out wrong, their fear of secularism and evolution is a real emotion that is experienced in the same visceral way that our own fear is felt. While we can address the creationists' misconceptions about the natural world and attempt to correct it, we can simultaneously show compassion and empathy for the feelings that this awakes in them.

I cannot emphasize enough here the need to differentiate between the creationist movement which poses serious threats to our science education system and a creationist as a person who is simply a product of this movement. By treating this person with the respect that every human deserves and communicating compassionately, instead of mocking them in a Dawkins-styled approach, they may feel safe enough to become willing to listen and engage with the data presented to them. This safe space is not a luxury but a necessity that creationists require in order to be able to begin question their previous conceptual structures (misconceptions) and to move towards an understanding of the validity of the theory of evolution. We must understand that the fate of the conceptual seeds that we plant rests largely on the means in which these data are presented to them.

Validation thus in this sense is simply the recognition of a person's value as a fellow human being and an empathetic acknowledgement of their emotional experience. We do not have to agree with someone's beliefs, but we can accept that those beliefs are important to them and that their fears are real. Moreover, from a compassionate standpoint, we can also acknowledge that although we do not share their fears, their fears remain as real to them and cause them the same pain that our own fears cause us.

Yet validating someone else's emotional experience is not easy when we are caught up in our own. In fact, studies have shown that individuals are more likely to show empathy and compassion to a person who is suffering once that person has been able to manage their own emotional distress (Jennings and Greenberg 2011). Thus, it is important that educators also validate themselves and their own feelings by acknowledging how difficult it is to teach and discuss evolution with those who are fundamentally resistant. This is especially true in formal education settings since teaching is an emotionally laborious job even on the good days. While teachers are not necessarily expecting students to be openly grateful, there is a certain desire and expectation of receptivity. It is therefore understandably frustrating, even maddening, when one is instead met by students who vigorously challenge everything put forth in a well-planned lesson.

It is therefore vital that these teachers refrain from judging themselves harshly when these negative feelings arise, and instead recognize that it is natural to feel rundown and frustrated in such difficult situations. When we combine radical acceptance and self-validation in the classroom, we see a teacher who accepts that she will have creationist students in her classroom and when she becomes frustrated by the situation, she reminds herself that this type of frustration is natural when you have dedicated your life to teaching and the students refuse to learn. But this self-validation should not be confused with self-righteousness. It is just a kind stance towards oneself and one's emotional experiences, acknowledging that it is natural that these feelings have arisen given the present conditions, but that they will also dissipate naturally if we do not perpetuate them with negative judgements. The same is true for all of those who engage in informal education. Later in this chapter, we will discuss specific breathing exercises that can also be used to reduce emotional turmoil and help regulate extreme emotional states.

## Classroom and Communication Reflections

As long as we are running on autopilot, we are often unaware of our thoughts and how these thoughts affect our moods and our responses to people around us. This autopilot mode means that we tend to react to situations based on long established behavioral patterns. The most exciting aspect of mindfulness is the ability to recognize this mode and to step out of it. As we continue to take on the role of the conscious observer, we are better able to rationally choose our responses. In this way, we take advantage of the brain's plastic nature and begin to create new neural networking that subsequently support our ability to shift gears and make conscious, rational action choices in the future (Figure 21).

The well-known phrase, 'neurons that fire together, wire together' describes our brain's ability to 'learn' certain behavioral patterns, i.e., the more often we engage in certain behavioral patterns, the more these behavioral patterns are enforced. In this way, the brain is trained to react to similar stimuli even faster and more effectively when we are exposed to similar situations or conditions in the future. This type of neurological wiring process is very advantageous when we are trying to perfect a certain behavior such as a particular type of sport or musical instrument. It is however disadvantageous in terms of unhealthy or unwanted behavioral processes or reactionary tendencies.

To understand how this type of wiring affects science education and science communication, let us imagine a person who is convinced that evolution is a dangerous idea and that 'believing' evolution causes a loss of faith and morality. Each time this person encounters another individual who attempts to explain the data that support the theory of evolution, this person automatically becomes tearful and defensive, immediately arming themselves with counterarguments and accusations. The more this person engages in this activity, the quicker they are to march down the proverbial war path as soon as someone mentions evolution. This example can also be seen from the opposite point of view, if we imagine a person who is convinced that all creationists are ignorant fundamentalists. Based on these judgements, this person is likely to act in disgust as soon as someone says they doubt that evolution is true. A productive, sensible dialogue concerning the validity of the theory of evolution would be impossible between two such individuals and a confrontation between two such individuals will most likely lead both individuals to feel justified in maintaining their original belief system, prejudices and reactions. Thus, an axiological shift in perceptions and behavior is necessary for these two individuals to be able to have an effective and educational conversation.

This is where self-directed neuroplasticity or conscious evolution becomes incredibly interesting because we can undo unhealthy or unwanted behavioral patterns and support the development of healthy or beneficial habits through conscious choice. The first step towards this type of conscious evolution is awareness. When a person becomes aware of their behavioral tendencies, they slowly gain the ability to choose how they want to act in a particular situation. This is particularly true when addressing conceptual conservatism. We have already discussed the multilayered socio-psychological reasons for creationist clinging and how our evolutionary past is embedded in the primitive parts of the brain causing a drive to avoid pain, seek pleasure and uphold familial bonds. Despite the many reasons for why creationists cling to their beliefs, there are also many reasons for them to loosen their grasp on these beliefs as these belief systems can be very limiting, not only educationally but also socially and psychologically. To be able to loosen their grasp on these dogmatic beliefs, the creationist first needs (1) to recognize that they are clinging, and (2) to realize that this clinging is a choice, i.e., optional.

While this type of awareness can be developed through a number of formal and informal practices, I have decided to design as heuristic based on the work by Tobin's group at CUNY which has already proven to be effective in formal education settings. I have created a compilation of those characteristics that I believe are most relevant for discussions involving human origins and adapted a number of them to make them more specific to communication on evolution (Table 13). While some of these points are specific for use in public-school classrooms, many of these characteristics can support effective dialogues outside of an educational setting. In that case, the word 'classmate' could easily be replaced with audience or conversation partner to highlight this applicability.

The components of this heuristic are divided into three sections as a reflection of the major components of the conflict. In other words, it is necessary that (1) students reflect upon the general guidelines and dynamics of the classroom. The clear separation of church and state in America means that teachers cannot discuss creationism or intelligent design in the classroom, nor can it be taught parallel to evolution. It is also imperative that (2) students become aware of their own internal world of thoughts and emotions and make the attempt to engage with their upstairs brain. Equally important is that (3) students and teachers become aware of how their actions and words affect others and the general classroom environment. Here I was sure to integrate components that highlighted the value of validating others' experiences and listening to learn from one another.

The goal of the *Classroom Guidelines for Respectful Communication and Interactions* (Table 11) and the *Talking about Human Origins Heuristic* (Table 13) is to create learning environments where students feel safe. The whole point of this is not to encourage creationist students (or teachers) to use this space to 'teach both sides', but instead to prevent a fight-or-flight response or a boomerang effect. Creationist students will often enter the classroom already seeing evolution as a threat. Aggressive speech or belittlement in the classroom will only reinforce this belief. Therefore, it is essential that we try to maintain positive, compassionate classroom environments as we have already discussed how the overall emotional climate of a classroom can have a severe impact on learning outcomes. Depending on the classroom dynamics, teachers can choose to incorporate additional characteristics such as 'I understand the difference between faith and science' or to remove certain

**Table 13:** Proposed heuristic for discussing human origins in the classroom with focus on classroom dynamics, awareness of emotions and interpersonal relationships.

| Talking about Human Origins Heuristic |
|---|
| *When discussing conflicting beliefs about human origins* |

*Awareness of classroom guidelines and dynamics:*

1  I realize that neither creationism, creation science nor intelligent design can be taught in a public-school science course due to the separation of church and state as mandated by the U.S. Constitution.

2  I am aware that people who have strong religious backgrounds or literal beliefs about the Bible may be more sensitive or feel hurt when discussing evolution.

3  I am aware that people without religious backgrounds or with religious beliefs different than my own may feel uncomfortable or offended by religious conversations or religiously-based arguments against evolution.

*Awareness of opinions and their effect on emotions:*

4  I am aware of when I identify with my opinions, thoughts and views, believing them to be the only correct view of reality.

5  I am aware of when I judge another person for holding an opinion or view in conflict with my own.

6  I am aware of my desire to convince others that their viewpoint is wrong and my attempt to convince them that my viewpoint is correct.

7  I recognize the negative emotions associated with clinging to a particular view or trying to defend a particular opinion.

8  I recognize when others suffer from negative emotional states when they attempt to prove themselves right or defend a particular opinion.

9  I try not to get disproportionately stuck when I am offended by opinions by others.

10  I am aware when I am overcome by very intense emotional states during classroom discussions and how these states affect my thoughts and perceptions.

11  When I am overcome by intense emotions, I try to focus on steadying my breath or engage in other breathing exercises.

12  I attempt to acknowledge deep emotional challenges without getting stuck.

13  I show myself compassion when appropriate.

*Awareness of interpersonal relationships within the classroom:*

14  I respect everyone's identity and background, without assuming anyone's beliefs, worldviews or perceptions to be the same as mine own.

15  I acknowledge and respect views, values, knowledge systems, and life histories of others, even when they may differ from my own.

16  In the process of addressing and discussing different worldviews, I avoid a patronizing stance, stereotypes or acting condescendingly.

17  I listen with respect, even when I do not agree.

18  Even when I experience emotional discomfort, my speech remains intentional, meaningful, honest and kind.

19  When others possess different worldviews, belief systems or identities, I use the opportunity to validate their experiences and perspectives.

20  Through my talk or actions, I make sure I show solidarity and kindness to those who experiencing pain, insecurity or fear when discussing organic evolution.

21  I structure my talk and actions to discourage stigmas and feelings of "not belonging".

22  I try to understand my classmates by listening and learning how their life experiences differ from my own.

23  I use my speech and body language to validate my classmates' experiences and perspectives, even when they are very different than my own.

24  I refrain from stereotyping and condescending when speaking.

*Table 13 contd. ...*

*...Table 13 contd.*

| Talking about Human Origins Heuristics |
|---|
| *When discussing conflicting beliefs about human origins* |
| 25   I try to sustain difficult conversations with the intention to understand, even when these conversations get uncomfortable. |
| 26   My classmate and I trust one another to listen to one another's opinions and beliefs without thinking badly of one another. |
| 27   I forgive those who may offend me during respectful and well-meaning conversations related to evolution. |
| 28   When I have offended someone, I try to become aware of the transgression and any harm done, and if necessary, accept responsibility. |
| 29   I have understanding for my classmates and teacher when they are overcome by intense emotions, even when it causes them to overreact in during discussions. |
| 30   My classmates, our teacher and I attempt to find common ground to re-establish our connection after disagreements. |

characteristics. The heuristic here is only to act as a guide for possible classroom implementations. While the introductions of heuristics and classroom guidelines are two means of developing a classroom-based mindfulness intervention, another means of reducing stress, developing greater awareness and fostering the development of trusting relationships is through the use of breathing exercises.

## Breathing to Reduce the Fight-or-Flight Response

Breathing exercises can be part of a formal meditation practice or they can be used intermittently throughout our day to ground ourselves in the present moment and relieve stress. When it comes to classroom purposes, long sitting mediations where students observe the ebb and flow of their breath are not particularly practical. While longer breathing exercises could be more easily introduced into a health class curriculum, short and simple breathing techniques are easier to implement in traditional classroom settings as these types of simple breathing exercises could be taught to students and used in the class without detracting from classroom time.

Looking back at the work of the CUNY group, we see that Alexakos et al. stated that they would begin each of their classes with a simple three-minute breathing exercise (2016). While they did not specify exactly what type of exercise was done during this three-minute period, they do refer to the physiological benefits of deep abdominal breathing because it supplies the body with a greater percentage of oxygen than a shallow breath. They also write about how the deep abdominal breathing supported the students in becoming more aware of their physio-emotional state and thus represented a high-grade intervention linked to mindfulness that assisted in ameliorating negative emotional states during dialogues related to thorny issues (Alexakos et al. 2016). They also attributed this deep abdominal breathing with more positive emotional and physiological states—both of which have obvious implications for better learning environments and more effective communication.

The calming effect of breathing exercises could be key to preventing a fight-or-flight in response to lessons on evolution because the fight-or-flight response is linked to certain bodily responses that are initiated automatically for survival purposes via the autonomic nervous system. In addition to the familiar physical symptoms of a threat perception—e.g., increased heart rate—there are also significant cognitive effects of the fight-or-flight response. The freeze-or-faint side of the fight-or-flight response results in cognitive shut down, for example. As long as a person is involved in this stress response, learning is greatly impeded—if possible, at all. So, it is imperative that this physiological response is mitigated or avoided entirely in order to keep students cognitively available.

So, what can we do to prevent or counter this fight-or-flight response? Is it really as easy as taking a deep breath? In fact, it might just be that simple due to the intriguing wiring of the autonomic and somatic nervous system. First and foremost, we must understand a vital point, one that we are all aware of, but may not truly understand the profound implications of: Our breath is connected to both the autonomic (involuntary or visceral, formerly vegetative) and somatic (voluntary) nervous systems.[22] In other words, breathing occurs automatically without having to actively think about it, yet we also have the possibility to actively control the breath if desired. The respiratory system is unique in this way as it is the only system that is connected to both the voluntary (somatic) and involuntary (autonomic) nervous systems.

To better comprehend how intentional breathing can act as a gear-shift, it is important to understand the sub-division of the involuntary autonomic system into two sub-systems, the parasympathetic and sympathetic systems. In the simplest of terms, the sympathetic system and parasympathetic systems are responsible for excitation and inhibition respectively. In other words, the sympathetic system is closely associated with the fight-or-flight response and readies the body to respond to a threat. When the sympathetic system is active, excitatory neurotransmitters cause your heart to pump harder and quicker, your peripheral blood vessels to narrow and tighten. On the other hand, when the parasympathetic system is active, it causes your heart rate to slow and peripheral blood vessels to widen. In an ideal situation, these two systems work in coordination to ensure that the body responds appropriately to different situations. In other words, in a dangerous situation blood should be available in the skeletal muscle for quick responses. Yet in the absence of danger, blood should remain around the major organs to support them in their function, e.g., digestion. Thus, the two portions of the autonomic systems can be extremely oversimplified by saying that sympathetic system is responsible for fight-or-flight response and parasympathetic is linked to a state known as rest-and-digest. We will concentrate here on this subdivision of the autonomic system in order to discuss how the somatic system can be used to shift gears within the autonomic system from the excitatory sympathetic system to the relaxing parasympathetic system.

The key to this shift is the vagus nerve. The vagus nerve can be understood as the major highway that allows voluntary input in the form of intentional deep-

---

[22]    While the somatic nervous system is responsible for controlling all skeletal muscle, the autonomic system regulates visceral processes in the body through the innervation of smooth muscle, cardiac muscle and glands. In vertebrates, the voluntary system only involves excitation, while the involuntary system involves both excitation and inhibition.

breathing from the somatic system to affect the functioning of the involuntary autonomic system. The vagus nerve is the longest nerve of the autonomic nervous system in the human body with both parasympathetic and sympathetic components. Its parasympathetic responsibilities affect the heart, lungs, and digestive tract. Research has shown that breathing deeply and slowly into the abdomen stimulates the parasympathetic portion of the vagus nerve and can bring about an almost instant reduction of the sympathetic fight-or-flight responses in the body (Wang et al. 2010). It has also been found that this type of deep breathing also improves heart rate variability (HRV) which is associated with increased psychological and physical resilience, as well as better cognitive function (Lehrer and Gevirtz 2014).

What this means is that although the fight-or-flight response is governed by the *involuntary* autonomic system, it is possible to influence it almost instantaneously using the *voluntary* somatic system to guide conscious breathing, thereby shifting gears within the autonomic system from the sympathetic system to the parasympathetic system (Figure 23).

The beauty of these breathing techniques is that they act as a means for us to consciously and actively manipulate our physiological systems, thereby influencing our psychological perceptions. In other words, intentional breathing allows us to take advantage of the neurological link between body and mind. While most people are aware of the brain's ability to recognize a potential danger and cause a physiological response based on this perception, fewer individuals realize that this same type of body-mind communication is possible in the other direction. In other words, through controlled body movements and breathing, the body can also 'talk' to the brain and convince it that it is 'safe' and that it is okay to 'relax'. While we may not all be aware of this neurological two-way street, most of us have experienced this phenomenon through engaging in sports, either by going for a walk or a run or by participating in a yoga class. The proverbial ability to 'clear our minds' by going for a walk is caused by this two-way connection between the body and mind.

This idea was reflected in the heuristics developed by the CUNY group (Table 10) and in the work by Immordino-Yang and Damasio which showed that

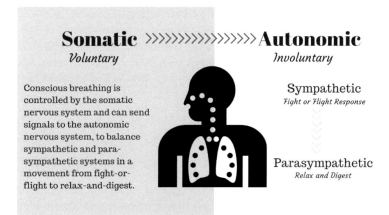

**Figure 23:** Overview of the autonomic and parasympathetic control over respiration.

the bodily sensations of physical changes caused by the tensing or relaxing of skeletal musculature or changes in heart rate contribute to changes in emotional states and indirectly influence thought patterns (2007). This ability to use conscious physiological processes to override maladaptive survival mechanisms is crucial in the movement away from primitive behavioral tendencies. Despite having ancient survival responses hardwired into our basic neurological networks, our more highly evolved neocortex allows us to consciously override this system in order to gain the necessary time and space to rationally decide how we want to respond to stimuli in our environment. In other words, while stressful classroom situations may send students (or the teacher) into a major fight-or-flight response, breathing exercises allow us to counteract this evolutionary artefact in order to quiet the mind.

So, while the proposition that breathing exercises can support more effective science learning may not have seemed like an obvious solution initially, once we have understood how emotions affect and hinder learning, it becomes clear that we need tools to alleviate the effect of intense, negative emotions on cognitive functions. While we are unable to remove the emotionality from a creationist's view of human origins, we can offer them tools to recognize and regulate the emotional responses that this topic cause within them because ultimately, all forms of teaching are useless if students are cognitively unavailable. So, we can now see how deep breathing supports better cognitive functioning and thus it becomes clear that mindful breathing exercises represent a new and potentially vital tool in supporting classroom learning.

Thus, we can see that breathing exercises can have a profound effect on the autonomic nervous system and cause a major shift from sympathetic activation to parasympathetic activation—basically turning off the fight-or-flight response. By moving the students out of the fight-or-flight mode, we are not only supporting their physical and emotional well-being but also opening the door to greater cognitive flexibility. These types of breathing exercises could be implemented school-wide, since we all know that discussions involving evolution are not the only emotional triggers in students' lives. Yet, in most parts of the world, we are still very far from such initiatives. For that reason, I have put together a brief overview of three exercises that could be implemented in a biology class without detracting from lesson time. The initial introduction of these exercises could be done during anatomy and physiology lessons that discuss the respiratory system or nervous system or the fight-or-flight mechanisms, thereby actually reinforcing the taught materials. Once the students are familiar with these exercises, they can be implemented throughout the school year in 3–5-minute intervals or simply referenced in a class heuristic or guideline. The main point is that students become aware of such exercises and have a precursory understanding of how they work. An interest and utilization of these exercises, like the understanding of evolution, may not occur until many years after classroom instruction, but the most important point is that the seed is planted.

I have included a full description of three potential exercises in Appendix III. While all three of the exercises have different focuses, they are all centered around an observation of the breath. For the first two exercises, the breath acts as an anchor for the attention. By trying to follow the breath, one becomes automatically more aware of the mind's occupation with thought, as Gil Fronsdal states, "Even when you have trouble staying with the breath, your continued effort to come back to the breath can highlight what might otherwise be unnoticed, i.e., the rapid momentum of the mind. In fact, the faster our thinking and the greater the preoccupation, the greater the need for something close by like the breath to help bring an awareness of what is going on. That awareness, in turn, often brings some freedom from the preoccupation (2008, p. 20)."

Breaking through this preoccupation, by using these types of breathing exercises that focus on the breath and ground us in the present, allows practitioners to gain access to a quieter mind. According to Fronsdal, one of the greatest benefits of these types of practices is this increasing familiarity of a state of calm and inner-peace. Knowing this state of being allows individuals to more easily recognize behaviors, thoughts, judgments that detract from this inner sense of peace, and the more one becomes acquainted with feelings of well-being, the less willing we are to sacrifice this inner peace, for instance by clinging to negative judgements or engaging in aggressive debates.

Of these two breath-observation exercises, I believe that the most crucial and effective exercise is the *Conscious Observer* (Figure 29, Appendix III) as it combines the principle of mindfulness of thought with intentional breathing. This exercise gives practitioners the insight into their higher cognitive functions and allows them to better recognize thoughts and emotions as products of the mind and their circumstances and not identify with them wholeheartedly. It allows them to move into the 'observer' position, as this allows the practitioner to gain a certain degree of cognitive distance from their thoughts and emotions in order to rationally decide if they want to pursue a certain train of thought or if them want to act upon a certain impulse or emotion. Again, I believe the best way of visualizing this is that the role of the 'observer' is the activation of the upstairs brain, which then 'observes' the activity of the downstairs brain. This movement towards upstairs functioning is greatly supported through the physiological shift in the autonomic system through the conscious breathing. This physiological shift is even greater in the third exercise which consists of an intentional manipulation of the breathing pattern.

The description of the second two breathing exercises are just a few of very many possible practices. This should not be understood as a comprehensive overview but is meant simply as a glimpse into the myriad of practices available (Figure 30, Appendix III). As we can see from the first of these two exercises, the main focus here is again the ability to become aware of the presence of thoughts and feelings and the ability to allow them to arise and pass away without impulsively reacting to them. Of course, to make these exercises more effective, I would encourage teachers to amend these exercises to make them more specific to their classroom needs, e.g., "Observe your thoughts, judgements and emotions regarding human origins" (Figure 30). As just stated, the last exercise is a clear example of how the breath can be used to shift from the sympathetic to parasympathetic system by sending impulses to the brain that it is okay to relax (Figure 23). Again, this is just a brief glance at mindful breathing exercises, for those interested in finding out more, there are a number of books devoted entirely to breathing exercises, such as *The Art of Breathing: The Secret to Living Mindfully* by Danny Penman.

The purpose of the classroom reflections and these breathing exercises is to offer students and teachers or creationists and new atheists a look into the inner-workings of their mind. By recognizing our tendency to judge or to cling to particular opinions, it is easier for us to recognize how we quickly become engaged in heated discussions over topics such as human origins. This awareness also allows us to recognize the possible means that we can avoid such divisive behavior by regulating our emotions in an effort to respond rationally and effectively. The development of this emotional awareness as well as the skills necessary to regulate emotions supports the creation of positive communication dynamics that foster learning and compassion instead of close-mindedness and division.

These breathing exercises can also help alleviate the negative emotional states that might arise in creationists once they do begin to question the validity of the special creation and move towards an acceptance of evolution. As discussed, the level of conceptual change that is required for a young-earth creationist to come to terms with evolution means that there is a high chance that this restructuring of conceptual structures will also negatively affect existing meaning systems. In order for true learning to occur, though, it is not enough to only address negative emotions and the need to prevent a fight-or-flight response but it is also imperative that we address how positive emotions can be produced that motivate these individuals to *want* to learn about evolution. In the next section we will look at the possibility of motivational gear-shifting within an educational context.

## Shifting Gears from Threat-Recognition to Goal-Based Learning

From all of the points that we have discussed so far, from the legal strife to the psychological complexities of creationist clinging, it is beyond doubt that teaching evolution can be a daunting task, particularly in the face of overt rejection and denial. The emotional impact of the topic, coupled with the societal division and the involvement of think-tanks, means that lessons on evolution may appear to be more of a battle than a typical classroom exercise.

In the past chapters we have already discussed how this division and aggression feeds into creationists' fears of secularism and scientific advancement. Their desire to cling to and to defend creationist beliefs in face of contrary data is grounded in the ties between these beliefs and their self-theory and world-theory. Doubting these beliefs can lead to a destabilization of these theories and thus affect the way that the creationist functions in the world and their ability to handle adversity. For the creationist, secularism and evolution are perceived as a threat and this perception can cause them to shut down or to become defensive within formal education settings. While we cannot remove this fear, it is possible to create educational environments that do not *reinforce* this fear. We can imagine that if a creationist student is ridiculed in science class that this student will feel vindicated in their fear of secularists, non-Christians and science. This is why pro-social classrooms that promote acceptance and inclusion are necessary to prevent such situations from occurring and why it is necessary to provide both teachers and students with the necessary tools to regulate intense, negative emotional states. Once the emotional side of this issue has been addressed, we can finally move on to the intellectual side and begin to look at how we can increase curiosity and scientific literacy.

As we discussed in earlier, there are three primary motivation systems: (1) threat system or fear-based, (2) drive system or goal-seeking and (3) befriend system (Table 5). It is not enough to avoid triggering a fight-or-flight response, instead we must also actively focus on activating those systems that support both learning outcomes and social dynamics, namely our drive system and our befriend system.

As discussed, these intrinsic motivation systems can be activated by extrinsic factors such as the use of sticks (punishment) and carrots (reward). These external impulses work because they correspond to internal motivation systems that are related to our most primal pleasure-seeking and pain-aversion instincts. When talking about motivation systems, though, they receive new titles such as the threat-aversion (fear-based) motivation and reward-seeking (goal-based) motivation (Table 5). To understand how these work, we can take a simple example, like a student's motivation to study for an exam: Some students may study because they are motivated by fear, i.e., the fear of failing an upcoming exam, while others engage in the same behavior—studying for the exam—but they are motivated by goal-oriented behavior, e.g., in the desire to get a good grade in order to attend a top university.

Without these intrinsic motivation systems, the external systems would have no effect on the student—in other words, the carrot activates intrinsic goal-seeking motivation systems while the stick activates intrinsic fear-aversion motivation systems. While both the stick and the carrot can both function as a means of motivating students, the activation of the fear-aversion system comes with a battery of neurological and emotional side-effects that negatively affect learning and the overall well-being of a student.

We have seen that creationist belief perseverance can be explained through an intrinsic desire to avoid pain and seek pleasure. Fundamentalist teachings extrinsically reinforce this type of clinging by actively teaching adherents of the mortal dangers of accepting evolution. The resulting motivation to reject evolution can become so high that students will intentionally fail tests on evolution as a form of martyrdom, as will we see in the next chapter (See *Conversations with Pete*).

So what kind of educational strategies can be developed to encourage interest in science when we have students entering the classroom who are intrinsically and extrinsically motivated to reject lessons on evolution. As we know, the threat system is often capable of overriding the drive system and the befriend system. In the absence of positive stimuli to trigger either the drive or befriend system, we can expect that the creationists' fear system will continue to dominate and lead to further rejection of evolution and general cognitive shut-down. It is therefore imperative that we develop positive learning experiences that encourage the activation of either the drive or the befriend system.

Research on the relationship between learning and the brain has shown that intrinsic motivation increases when students have a sense of autonomy, in other words, when they have control in decision-making and self-determination (Behar-Horenstein 2006). This is particularly true during adolescence when teenagers are undergoing a major milestone in developing their independence from their parents. Here it is important to remember that as part of development of autonomy, adolescents make a shift away from seeking praise from authority figures and attempt instead to obtain approval and acceptance from their peers as Behar-Horenstein stated, "Adolescents are motivated

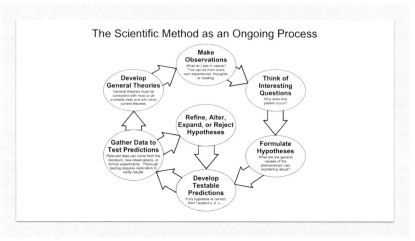

**Figure 24:**  Overview of the scientific method (Image: ArchonMagnus from Wikimedia Commons).

by social goals as much if not more than academic goals so that the adolescent culture creates their cognition. Adolescents' need for inclusion and avoidance of exclusion motivates their behavior toward school, activities, and learning. These principles dramatically alter how teachers motivate adolescents versus motivating younger students (2006, pp. 310–311)". Thus, the best means of enhancing intrinsic motivation is through "activities that enhance perceived competence and are optimally challenging (Behar-Horenstein 2006, p. 312)", while also incorporating cooperative collaboration to address the adolescent focus on group inclusion.

Incorporating what we know about motivation and learning as well as the changes in adolescent neurological structures, I have come up with two general lessons that focus on group cooperation, while also facilitating student-guided inquiry in an attempt to solve complex puzzles. These exercises can be incorporated into any lesson to increase interest in evolution or they can be used when a teacher is directly challenged by anti-evolution materials or accusations. The ultimate goal is to refocus students' attention on science learning by activating their goal-seeking system in an attempt to solve the problem at hand while also triggering the activation of the befriend system through cooperative group work. Ideally, both of these exercises should develop scientific curiosity and literacy as they allow students to discover the validity of the theory of evolution for themselves. Before beginning either of these lessons, it is useful to remind students of the scientific method (or hypothetico-deductive method) and encourage them to use this method throughout both of the lessons (Figure 24).

## *Lesson: The Mystery of the Missing Chromosome*

In this lesson, students are introduced to the chromosomal aberration between great apes and humans. While great apes have 24 pairs of chromosomes, modern humans have only 23. The disappearance or deletion of a pair of chromosomes would be lethal, so it is up to the students to solve the mystery of the missing set of chromosomes. The basic idea behind this lesson is to have students come up with possible reasons or hypotheses that explain how this difference in the number of chromosomes could have occurred and then find hypothetical ways to 'test' their ideas. Some students might use this opportunity to cast doubt on evolution and propose that humans and primates do not in fact share a common ancestor. That is okay, as students should understand that science is not dogmatic by nature, but they need to come up with clear ways of 'testing' their hypothesis. Most students, though, will base their ideas upon their acceptance of evolution as a valid theory and come up with ways of explaining the deviation in the number of chromosomes, for instance that the common ancestor originally had 23 pairs and that one pair broke, creating an additional pair in subsequent hominids and great apes or that the original common ancestor had 24 pairs and that one of the pairs fused. Regardless of what types of ideas and hypotheses the students come up with, the teacher should guide their exploration of the topic, posing questions such as: How could that idea be tested? What sort of evidence would be found if your hypothesis is correct? What sort of evidence would disprove your hypothesis? These questions are particularly important for those students who propose that this deviation proves that great apes and humans do not share a common ancestor.

Once the students have come up with a hypothesis, it is important for them to come up with ways that they can test their ideas. Based on what they know about chromosomes, what would be the sign of chromosomes being fused? How could we test to see if chromosomes had split? The groups that proposed either a break of a fusion should come up with diagrams to explain their 'expected' findings. Obviously, students will not be able to carry out real experiments but luckily, this chromosomal mystery has actually already been solved. The detailed lesson plan in Appendix V goes over how teachers can reveal the answer to students using diagrams or a literature search.

The most important thing from students to take away from this thought experiment is a better understanding of how science works and how it has been possible to test and validate the theory of evolution using the scientific method. The theory of evolution was proposed decades before scientists were even aware of chromosomes, meaning that the discovery and advanced understanding of genetics gave scientists new opportunities to test Darwin's theory. Instead of refuting the theory of evolution, advances in genetics have only substantiated the theory of evolution further. By the conclusion of the lesson the students should realize that this chromosomal deviation could have gone either way—either proving or disproving the theory of evolution. For evolution, particularly the common descent of modern apes and humans, to be correct there *must* be some biological evidence to explain the variation in the number of chromosomes and this is what was in fact found as human chromosome two is clearly the product of a chromosomal fusion as it contains both vestigial telomeres and centromeres, as published in *Nature* in 2005, "Human chromosome 2 is unique to the human lineage in being the product of a head-to-head fusion of two intermediate-sized ancestral chromosomes (Hillier et al. 2005)."

## *Lesson: Testing the Theory of Evolution*

This lesson is designed to teach students how the scientific method can be used to test the theory of evolution with regard to transitional fossils. Here students will learn how scientists can develop and test hypotheses about the logical location of transitional fossils based on the theory of common descent with modification. The idea of 'missing links' is crucial to the creationism | evolution debate as creationists often claim that there is a lack of intermediate forms within the fossil record thereby insinuating that the theory of evolution must be incorrect, while design proponents make similar arguments using different terminology such as 'information problems' (see *Ten Questions* by Wells and Dembski). This lesson also provides students with a great overview of macroevolution, which has been argued to be one of the most effective ways of teaching any student or adult about the validity of the theory of evolution (Padian 2010). The most important point here for students to take away is that although we are not able to go back in time and observe evolution, it is possible to use the theory of evolution to make predictions and test those predictions about the location of so-called 'missing links' or transitional fossils.

The importance of these transitional fossils to the theory of evolution was recognized by Darwin himself as he developed his theory and stated that according to his theory, the fossil record should show "infinitely numerous transitional links" and yet this fossil evidence had not yet been found at the time that he developed his theory. This concerned Darwin as he stated, "Why then is not every geological formation and every stratum full of such intermediate links? Geology assuredly does not reveal any such finely graduated organic chain; and this, perhaps, is the gravest objection which can be urged against my theory (1859; p. 280)."

After Darwin published his theory in 1859, paleontologists went out in search of these fossils to test Darwin's new theory and to see if there was any fossil evidence that could explain the origin of birds. By 1860, a fossilized feather had been found in Germany in limestone dating back to the late Jurassic period. This feather was subsequently described as *Archaeopteryx lithographica* by Christian Erich Hermann von Meyer in 1861. Just two years later, Richard Owen described a nearly complete skeleton of *Archaeopteryx lithographica*, which included many reptilian features, such as clawed forelimbs and a bony tail, despite its overall similarity in appearance to a bird (1863). Thomas Henry Huxley recognized that Archaeopteryx was a transitional organism that counted as one of Darwin's proposed 'intermediate links' based on the fact that it lived 155–150 million years ago and possessed both the characteristics of a bird and a reptile as he stated, "I think I have shown cause for the assertion that the facts of paleontology, so far as birds and reptiles are concerned, are not opposed to the doctrine of evolution, but, on the contrary, are quite such as that doctrine would lead us to expect; for they enable us to form a conception of the manner in which birds may have been evolved from reptiles, and thereby justify us in maintaining the superiority of the hypothesis that birds have been so originated to all hypotheses which are devoid of an equivalent basis of fact (1868, p. 75)." By 1870, the iconic 'Berlin specimen' of *Archaeopteryx*, had been discovered and provided further evidence of its role as a transitional fossil due to the presence of a complete set of reptilian teeth. This rapid succession of discoveries that quickly substantiated Darwin's theory may seem serendipitous to some, but they are actually examples of how the theory of evolution can be tested using the hypothetico-deductive method. In this proposed lesson (full description in Appendix V) students learn how to create

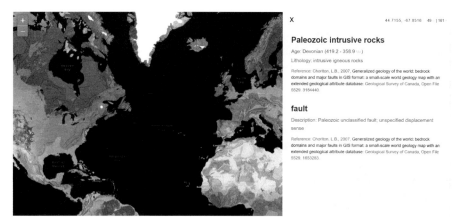

**Figure 25:** Example of interactive geologic map where students could find potential dig sites for Devonian fossil search.

**Color version at the end of the book**

hypotheses and 'test' the theory of evolution using phylogenetic trees, geological time scales, and an interactive geologic map of the world (Figure 25).

Like in the previous lesson, students are again unable to conduct the actual research but once students have come up with their hypothesized time of transition and place where they could dig for fossils, teachers can go over the story of a research group who actually did conduct this exact type of experiment: In 1999, Edward Daeschler, Neil Shubin and Farish Jenkins began to actively search for a particular fossil based on evolutionary and geological data. Using the hypothetico-deductive method, Daeschler and his team made a prediction based on the evolutionary theory. If fish gave rise to land animals then there should be evidence of an organism that shows both fish and amphibian characteristics and according to evolutionary theory, such an organism should have lived between 360 and 380 million years ago. So, it should be possible to analyze rock from this period of time and find evidence of these transitional organisms. In order to test their hypothesis, these scientists travelled to the Canadian arctic where large amounts of stone from an ancient Devonian shore line are exposed and there, they began to search for this fossil in order to test their prediction (and ultimately the theory of evolution).

In 2004, their efforts were rewarded with the discovery of what would later be called Tiktaalik on Ellesmere Island in Nunavut, Canada. According to Shubin, their expedition illustrates how the scientific method can be applied to evolutionary science and that because Tiktaalik has both the characteristics of primitive fish and amphibians it fills in the gap between water dwellers and land dwelling tetrapods (2009).

Both of these lessons could be expanded upon, amended and perfected by interested teachers or textbook authors. The generalized overviews offered here are only to serve as stimulus for future creativity and a humble example of a possible means of inspiring students through tasks that are challenging and increase students' perceived competence. These sorts of challenging, competence building exercises are optimal for increasing intrinsic motivation (Behar-Horenstein 2006)—particularly when done in collaboration with other students. Moreover, these lessons provide students with a sense of autonomy because they allow students to make their own decisions and develop their own self-determined approaches to the tasks, which also leads to increased levels of intrinsic motivation (Behar-Horenstein 2006).

Moreover, it is possible to adapt all of the information here in these lessons to use for general science communication in order to better describe how scientists have developed such a high degree of certainty in the theory of evolution since Darwin since proposed his ideas over one hundred years ago. In the end, the point is that these lessons or any communication about evolution should aim to awake positive emotions in connection with evolution such as curiosity, wonder and interest.

## Summary

The set of tools discussed in this chapter included: (1) classroom guidelines that act as a basis for respectful dialogues, (2) specific mindfulness practices that allow support parties on both sides to become more aware of their thoughts, emotions, options and relationships during the discussion of human origins, (3) specific stress-relief tactics such as breathing exercises that can be used to support mindful awareness and subsequently respectful dialogues and lastly, (4) lessons designed to shift gears from fear-aversion to goal-seeking behavior. While many of the mindfulness-based strategies may seem far removed from evolution education, they form an essential foundation for subsequent teaching and communication because they address the emotional factors that prevent learning and effective dialogues from occurring. While many specialists would like to focus solely on lesson content, we now understand the essential role of emotions and motivation in the learning process and can no longer afford to neglect this vital piece of the educational puzzle. The main hinderance when it comes to creationism is the intense emotional devotion that these students have to their creationist beliefs and the choice that they make to reject evolution in order to actively uphold their literalist beliefs.

That is why simply offering better diagrams, or clearer explanations of evolutionary theory is simply insufficient in addressing the psychological aspects of the creationism | evolution conflict. That is why I have placed so much emphasis on developing cognitive abilities in order to focus on disengaging the automated structures that hinder learning and ultimately conceptual change. The ideas presented in this chapter are grounded in the concept of brain-based education where neuroscience data are used to better understand how the brain learns naturally, so that more efficient educational techniques can be developed that promote the brain's biologically driven learning framework (Ramakrishnan 2013). In essence, this biological approach to effective teaching means that we are using our knowledge of biology to find a better means of teaching biology.

# Chapter 8

# Understanding Common Descent

*"All of us are part of an old, old family. The roots of our family tree reach back millions of years to the beginning of life on earth. [...] We began as tiny round cells, and we've changed a lot since then. But we carry with us reminders of each step of our past. That's how it is with families. And ours goes back a long, long way."—Our Family Tree: An Evolution Story by Lisa Westberg Peters*

Creationists often claim that the naturalistic methods used to explain the origin of species rob us of our humanity and lead to a loss of morality and compassion. Yet, when I look at the basic tenet of evolution, i.e., the concept of common descent, what I see is actually one of the most inclusive and harmonious views on the world as the idea of common descent actually describes the intrinsic connection between all living beings. While this core message is often overlooked in adult communications, it is the central theme in most of the books written for children on the subject. After failing at my attempt to explain evolution to my four-year-old daughter, I purchased three books on evolution for young children: *Grandmother Fish* (by Jonathan Tweet and illustrated by Karen Lewis) and *Our Family Tree* (by Lisa Westberg Peters and illustrated by Lauren Stringer) and *Older Than the Stars* (by Karen C. Fox and illustrated by Nancy Davis). Each of these books describes the origin of human life from an evolutionary and scientific standpoint and each of them has a clear, overarching message—all life forms are connected through our common descent from more primitive life forms. This point was extended even farther to our connection to the planet in *Older Than the Stars* which explains the interconnection of all things through our molecular origin during the Big Bang. Here with prose that are reminiscent of *The Man Who Swallowed the Fly*, children learn:

> *These are the people just like you, who live with the plants and animals, too,*
>
> *That grow on the planet green and blue, that circles the sun, our daily view,*

*That was born from the dust, so old and new, thrown from the blast intense enough,*

*To hurl the atoms so strong and tough, that formed in the star of red-hot stuff,*

*That burst from the gas in a giant puff, that spun from the blocks,*

*That formed the bits, that were born in the bang, when the world began.*

Yet, somehow this message of common descent, of interconnectedness is often lost in the heated debates over evolution and creationism. The same can be said of the central concept of Christianity "Love thy neighbor", which was important enough be stated in the Gospel of Mark, Mathew and John, but apparently is also absent when it comes to developing strategies aimed at disrupting classroom instruction. As we have seen, it is easy to overlook how we are all connected due to primitive survival mechanism that are programmed to enable us to quickly discern between 'us' and 'them'. Yet, the plastic nature of our highly evolved portions of our brains allows us to consciously and intentionally restructure our neurological wiring and mental constructions (Figure 20), which means that we have the cognitive ability to overcome our primitive tendency to 'other' and instead to choose to actively seek out our common humanity. In other words, we can choose to see the creationist as a human being independent from their fundamentalist views and can encourage the creationist to see the evolutionary biologist independent from their secular views. If we can stop defining ourselves by our differences, we may be able to find enough common ground in order to engage in meaningful conversations.

The division between creationists and secular science is just part of a larger division within society. We are currently living in societies that are increasingly divided over differences in opinions, worldviews and political affiliations. If our goal is to move towards a more progressive and harmonious society, we cannot assume that this can be accomplished by getting the 'others' to give up their beliefs, but instead I believe that we must have both sides move towards greater tolerance, empathy and open-mindedness. The solution lies in the willingness to move from diatribes and debates to discourse and dialogue (Figure 26).

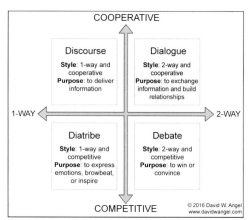

**Figure 26:** Communication forms—diatribe, debate, dialogue and discourse (reproduced here with permission from the author David W. Angel).

While our tribal past enables us to easily recognize our differences, we must make the conscious attempt to identify our common humanity in order to resolve this generalized societal riff. Failure to do so leads not only to ineffective teaching but also to unproductive science communication. It is not enough to present data and facts. We have to become more aware of how we are communicating with non-scientists and recognize the detrimental effects that certain forms of communication have.

One of the means of accomplishing this was discussed in the previous chapter regarding concrete classroom and communication practices that encourage both sides to reflect upon their own desire to uphold particular views. In this chapter we will look at further means of diffusing the situation. We will look at the need to drop aggressive, combative language in order to decrease division and conflict. Many of the so-called new atheists have not only labeled creationists as ignorant and insane but also made numerous offensive comments regarding religion in general. Because this type of contentious language only leads to greater division, it also increases the likelihood of a boomerang effect and conceptual conservatism. This clearly goes against our primary goal which is to use science communication and education to increase open-mindedness and cognitive flexibility. This mental flexibility is a prerequisite for subsequent conceptual change and ultimately the development of scientific literacy. This is by no means a small goal and entails a tremendous amount of effort from science communicators and teachers. It is not enough to demand that creationists become open-minded but we, too, must work on increasing our own cognitive flexibility in order to see beyond the simplistic black-or-white dimensions of this conflict. If we look at synonyms for open-minded, we find adjectives such as approachable, receptive, tolerant, observant, unbiased, impartial, perceptive, etc. While these are valuable characteristics for a student to possess, they also the key to becoming better educators and more effective communicators.

For this purpose, this chapter challenges new atheists' claims that all religion is detrimental to human development and that creationists are either evil, ignorant or insane. The proliferation of these statements only makes creationists feel vindicated in their fears of secularism and their felt sense of being under attack. A productive dialogue is necessary for conceptual change to occur. Insults and degradation of a system of belief that is sacred to an individual will never inspire that person to reconsider their conceptual structures or to adopt a more scientific understanding of the natural world. So, in order to help create a space for productive dialogue, it is not only necessary for creationists to drop their anti-scientific stance in order to be receptive to scientific lessons, but secularists will need to drop their anti-religion rhetoric so that they have a chance to awake interest in science instead of distain. We want to turn students *on* to science—but abusive language is always a turn *off*. For that reason, we will devote the next section to looking at the ineffectiveness of aggressive speech as a means of trying to convince creationists that they should become more accepting of evolution.

## *Preaching to the Choir*

It is beyond doubt that the desire to spread religious belief has led to horrible, bloody attempts to convert non-believers and it is understandable that the atrocious acts committed in the name of religion have caused a great deal of mistrust of religion in numerous individuals and inspired many of the comments made by the new atheists. As mentioned earlier in this book, the term 'new atheists' was coined by Gary Wolf in his 2006 article in *Wired Magazine* where he described the work of Richard Dawkins, Sam Harris, and Daniel Dennett. According to Wolff, "The New Atheists will not let us off the hook simply because we are not doctrinaire believers. They condemn not just belief in God but respect for belief in God. Religion is not only wrong; it's evil (Wolf 2006)." In essence, the new atheist are completely intolerant of religion and religious beliefs and believe that the faithful should be actively countered and criticized (Hopper 2006). While skepticism or mistrust is understandable to a certain degree due to the violence enacted in the name of religion, when this skepticism is taken too far, it quickly develops into its own dimension of closemindedness. This type of generalized attack on a group of people and beliefs is not only counterproductive, but it is the epitome of prejudice and intolerance. The situation is far too complex for this black-or-white type of thinking and a much higher degree of flexibility of mind can be expected of scientists and intellectuals.

This chapter examines two major new atheist claims, here epitomized in quotes by Steven Weinberg: "Anything that we scientists can do to weaken the hold of religion should be done and may in the end be our greatest contribution to civilization." and Richard Dawkins: "It is absolutely safe to say that if you meet somebody who claims not to believe in evolution, that person is ignorant, stupid or insane (or wicked, but I'd rather not consider that)."

I should first point out that while I respect Weinberg and Dawkins' passion for science and also share their concern regarding the anti-scientific, fundamentalist and literalist trends, I do not believe that their form of communication is effective in trying to reach their goal. In order for the general public and students to understand major scientific tenets, we need them to be receptive, we need them to be willing to tune in, yet these types of statements cause many to tune out, and even be turned off. This form of aggressive speech only puts creationists in the position where they feel more inclined to defend their views leading to belief perseverance. These aggressive attacks can therefore be understood as the quintessential 'preaching to the choir' as they only appeal to individuals who already share Dawkins or Weinberg's views. Such communication only serves to decrease scientific curiosity within those who are already weary of science.

Some new atheists such as Jerry Coyne have devoted entire books to promoting the idea that *all* forms of religion and faith are *always* detrimental. Anyone who has ever prepped for the SAT exams knows that any statement containing 'all' or 'always' is almost always incorrect because these statements cannot realistically represent the vast unpredictability and variation seen in the real world. These declarations are not only incorrect, but they also lead to great further division and increased intolerance. Moreover, these types of divisive statements are also entirely counterproductive to the authors' own intentions as *Scientific America* writer John Horgan pointed out in his review of Jerry Coyne's book *Faith vs. Fact: Why Science and Religion Are Incompatible.*

*Coyne's defenses of science and denunciations of religion are so relentlessly one-sided that they aroused my antipathy toward the former and sympathy toward the latter… . He overlooks any positive consequences of religion, such as its role in anti-slavery, civil-rights and anti-war movements. He inflates religion's contribution to public resistance toward vaccines, genetically modified food and human-induced global warming.*

While Coyne, Dawkins and other similarly minded individuals may offer many logical examples to support their opinion, it is not feasible, practical or constitutionally valid to expect (or even want) students to give up their religious beliefs. Of course, a key step towards scientific literacy is a student's ability to accept and understand scientific teachings, but this must be possible without going to battle with them over the meaning of life.

This point was addressed by Eugenie Scott, former director of National Center for Science Education, in her talk entitled "Equipping Scientists to Better Understand and Converse with Religious Communities" given at the 2014 American Association for the Advancement of Science (AAAS) convention. As Scott pointed out, the majority of American students are religious and "if students are presented with a dichotomous view—one must choose between science and religion, or evolution and creationism—the door effectively is shut to further scientific understanding (2014)." These type of anti-religious attitudes and attacks only serve to put religious students on the defensive, initiating a cognitive shut-down as they attempt to uphold their faith. The reasons for this active attempt to protect their belief system were discussed in length in previous chapters and here it suffices to say that telling students that they must decide between science and their faith will arguably result in a loss for science because those individuals who are truly convinced that their eternal life is dependent on their faith in their God would not sacrifice their faith solely for the purpose of appeasing their science teacher or passing a science exam. In fact, as we will see later in this chapter, some creationist students turn themselves into purposeful martyrs by intentionally failing exams on evolution (see *Conversations with Pete*).

While we cannot deny that having an outspoken creationist student in a science classroom makes teaching evolution more cumbersome, it is our responsibility as educators and as a society to educate all students, regardless of their socio-economic background, psychological disorders, intelligence, beliefs, etc. Certainly, teaching is always easier when students are open-minded, well-prepared and excited about learning, but this is not what we find in real classrooms—at least not the ones I have sat in as a student. Instead, there will always be students with attention deficit disorders, students battling addiction, students working over thirty hours a week to help support their families, students witnessing domestic violence on a weekly basis and, yes, fundamentalist students, who are convinced that science is wrong, perhaps even evil. We can spend our time wishing for a classroom full of easy, 'teachable' students or we can accept the reality as it is today and find a means of teaching and accepting all students.

We can, for instance, accept that we have religious students in a classroom and that for some of these students, their faith and religion play a central role in their lives and identity. We can acknowledge that this belief system will make it difficult for them to accept scientific theories, such as evolution, which are perceived as threats to their faith and we can still make a mindful attempt to help them overcome these fears in order to offer them the best education possible. But this will only be possible through discourse—not attack. As author and journalist Jeffrey Taylor points out, Coyne's book "*Faith Versus Fact* could serve as a primer for nonbelievers wishing to present their case to the faithful as well as an aid for doubters struggling to resolve theistic dilemmas themselves. Atheists might hope that it could challenge believers by picking apart arguments for religion's merits and veracity. But as his book demonstrates, and as the reactions to previous atheistic polemics by Sam Harris, Richard Dawkins, and the late Christopher Hitchens have proved, it's unlikely to dissuade those whose faith is strongly grounded. Science might be based on a foundation of rational thought and trial-and-error, but the roots of religion lie in something much more incalculable, and thus much harder to counter (Tayler 2015)."

As individuals like Coyne and Dawkins continue to preach to the proverbial choir, their combative words have no chance of effecting changes in the attitudes or views of the religious or fundamental. We will not be able to convince anyone to loosen their fundamentalist beliefs through such antagonist speech. Ultimately, it is almost comical to demand that the literalist to become more flexible in their thinking when one is not willing to question his own prejudices and mental inflexibility.

The prerequisite for a functional discourse is thus our ability and willingness to drop this 'us vs. them' attitude. Anyone who has ever been involved in a relationship is aware of the fact that we, as humans, are much more likely to listen and find the willingness to make compromises when we feel like we are valued and respected. Thus, in order to have any chance of undoing anti-scientific attitudes and increasing receptivity to lessons on evolution, we too much drop our rigid anti-religious thoughts. It is undoubtedly difficult to accept a student's religious beliefs when a person, such as Coyne, believes that faith is inherently "detrimental, even dangerous, and fundamentally incompatible with science (Tayler 2015)." Thus, in order to make it easier for anti-religious individuals to gain some acceptance for the role of religion of others' lives, we will briefly look at how religion and the belief in common myths have played a key role in human development, culture, education and ultimately in scientific advancement.

My goal is not to convince the new atheists that religion is entirely good or to suggest that it would be beneficial for all individuals to take on religious beliefs, but simply to call for more tolerance and open-mindedness. As renowned skeptic Michael Shermer states, "…for every one of these grand tragedies there are ten thousand acts of personal kindness and social good that go largely unreported in the history books or on the evening news. Religion, like all social institutions of such historical depth and cultural impact, cannot be reduced to an unambiguous good or evil (2006, p. 71)."

## Role of Religion in Human Development

Homo sapiens are unique in that they are the only species of animals that has ever been able to speak about ideas and things that do not exist. Moreover, humans are unique in that their belief in these cognitive constructs can become so strong that it causes them to alter their behavior, as historian Yuval Noah Harari writes, "You could never convince a monkey to give you a banana by promising him limitless bananas after death in monkey heaven (p. 24)." This type of blind faith is exactly what upsets and irks people like Dawkins and Coyne, but according to Harari there is also an evolutionary advantage to a shared belief in ideas or myths as "… fiction has enabled us not merely to imagine things, but to do so collectively. We can weave common myths such as the biblical creation story, the Dreamtime myths of Aboriginal Australians, and the nationalist myths of modern states. Such myths give Sapiens the unprecedented ability to cooperate flexibly in large numbers (p. 25)."

While it could be argued that this collective cooperative activity does not always have positive effects on human development, such is often the case in the promotion of nationalist myths, this general ability to cooperate flexibly was key for human development as it allowed individuals to gather in large masses around a common idea and work towards a perceived common goal. In fact, the relatively recent discovery of Gobekli Tepe (or Göbekli Tepe) has caused experts to entirely rethink the relationship between human development and religious practices. While it was once believed that religious practices developed *after* the rise of organized human cultures, the evidence found at Gobekli Tepe implies that human culture developed as a *result* of religious belief and practices.

Gobekli Tepe contains massive carved stone structures and is believed to be the world's first temple. The site is over 11,000 years old and the fact that hunters and gathers were apparently congregating this early in human history at a religious site has changed the way that experts see the role of sociocultural practices in human development. As Stanford University archeologist Ian Hodder points out, Gobekli Tepe changes everything (Symmes 2010). While it was long believed that humans first developed agriculture and settled communities and then began to construct temples, Schmidt now argues that it was the other way around, i.e., "the extensive, coordinated effort to build the monoliths literally laid the groundwork for the development of complex societies (Curry 2008)."

It is the immense size and scale of the stone constructions that support this claim as the construction of the seven-ton stone pillars would have required the coordination of hundreds of individuals, all of whom would have needed to be fed and housed during the endeavor—hence the eventual emergence of settled communities in the area around 10,000 years ago. What this means according to Hodder is that "…sociocultural changes come first, agriculture comes later (Curry 2008)."

Coyne may argue here that while this type of religious congregating may have played an important role in the establishment of complex Neolithic societies, it has not had any positive role in promoting human growth since then. Yet, if we look at the foundation of scientific revolution, we can see how religious interests also played an indirect role in the promotion of science and secular academic pursuits.

While the exact dates of the scientific revolution remain debatable, most scholars agree that the publication of Nicolaus Copernicus' *De revolutionibus orbium coelestium* (On the Revolutions of the Heavenly Spheres) in 1543 marks the beginning of the modern scientific movement, which was characterized by advances in mathematics, physics, astronomy, biology and chemistry and a transformation in societal views of nature. But what does this have to do with religion? It could easily be argued that there were two main developments that were necessary for Copernicus to have the influence that he did on the world and this is (1) access to education and (2) ability to publish a book that could be read by a mass of literate individuals. Now if we look at the roots of higher education in Europe and the development of mass literacy combined with access to printed information, we again find that these developments were largely driven by religious interests and religious organizations.

While early religions were often local and exclusive in nature, most modern religions are universal and missionary (Harari 2014).[23] This transition from local to universal religious belief in the first millennium BCE changed the course of human history as their emergence "made a vital contribution to the unification of humankind, much like the emergence of universal empires and universal money (Harari 2014, p. 210)." This new universal and missionary character of modern monotheistic religions led to the establishment of schools through the need to educate clergy who would be able to spread their monotheistic beliefs. By the early Middle Ages, monastic and cathedral schools were the most important institutions of higher learning in Europe (Ellison et al. 2009). Yet these religious institutions were not only devoted to religious education but also to the preservation and continuation of science learning as they kept the study of Aristotle and Plato alive well into the Middle Ages (Lindberg 2007).

As time progressed cathedral schools were established in major cities in France, Germany and Italy and slowly the first institutions that can be considered universities arose, which were originally centered around the teaching of theology, law and medicine (Ridder-Symoens 1992). Although the foundation of medieval universities is clearly rooted in religious organizations, once the universities were established, they did not restrict themselves to the study of religious texts, but instead coincided with the widespread reintroduction of Aristotle's work and soon thereafter Aristotelian texts as well as the study of the natural sciences soon became the center of university curriculum (Huff 2017). The importance of science at these early universities cannot be underestimated as Edward Grant argues that the "medieval university laid far greater emphasis on science than does its modern counterpart and descendent (1984, p. 68)." Toby Huff further argues that these medieval universities played a crucial role in the scientific revolution due to this focus on Aristotle and other scientific and philosophical texts, "Copernicus, Galileo, Tycho Brahe, Kepler, and Newton were all extraordinary products of the apparently procrustean and allegedly Scholastic universities of Europe... . Sociological and historical accounts of the role of the university as an institutional locus for science and as an incubator of scientific thought and arguments have been vastly understated (p. 344)."

Thus, we return to our point of origin: Copernicus and the initiation of the scientific revolution. As Dobrzycki and Hajdukiewicz point out, Copernicus' university education provided him with the necessary grounding in the mathematical astronomy as well as access to the philosophical and natural-science writings of Aristotle and Averroes. His education at Kraków University provided him with the theoretical knowledge and critical thinking skills, which allowed him to recognize the logical contradictions between Aristotle's theory of homocentric spheres

---

[23]    Yuval defines universal as a "superhuman order that is true always and everywhere" and missionary as the desire to spread this belief.

and Ptolemy's mechanism of eccentrics and epicycles, which led to the development of Copernicus' own theory regarding the structure of the universe (Dobrzycki and Hajdukiewicz 1969). Without this access to formal education and without the preservation of Aristotle's work, Copernicus may have never developed his ideas.

Copernicus was not only fortunate to have access to these texts and education, but also had the good fortune of publishing his *De revolutionibus orbium coelestium* in 1543, almost exactly 100 years after Johannes Gutenberg introduced the first movable type printing system in Europe and almost thirty years after Martin Luther published his famed 95 theses. While the development of the printing press led to the increased availability of affordable printed materials, the Reformation is credited with advancing literacy among the general population (Pettegree 2000).

Although the invention of the printing press was rooted in the desire to reproduce religious teachings, once the printing press was in existence, it transformed not only the religious world, but also assisted in furthering the scientific revolution. For several centuries, it was uncommon for private citizens to own books and if they owned any it was usually only the Bible, but along with the press came the access to affordable reading materials and there was subsequently a dramatic rise in adult literacy in Europe as the tradition of oral readings gave way to silent, private reading. The new availability changed not only the way that citizens began to learn and to be taught but also helped establish a scientific community within which researchers could easily share their work. It is for this reason that Elizabeth Eisenstein argues that the printing press was implicit in the onset of the scientific revolution as it brought about the idea of "cumulative and progressive knowledge" as data became part of a "permanent accumulation no longer subject to the cycle of rapid decay and loss (1980, p. 797)." Thus, we can see how the scientific revolution was supported by both the development of medieval universities, increased literacy and the printing press. While the impetus for these developments came from religious motivations grounded in the desire to spread universal, monotheistic belief systems, it also had the effect of providing access to formal education and ultimately brought about a cognitive revolution and a movement towards scientific scholarliness.

I have provided this brief description of how religious belief and communal worship led to cultural advances and how the movement towards universal religious belief indirectly led to increased literacy, access to printed materials and formal education not to argue that religion is always beneficial to society but simply to as response to the oversimplified portrayals of religion made by new atheists such as Jerry Coyne, who included the following quote from Robert Green Ingersoll (1833–1899) in his book 'Faith vs. Fact':

> *"We have already compared the benefits of theology and science. When the theologian governed the world, it was covered with huts and hovels for the many, palaces and cathedrals for the few. To nearly all the children of men, reading and writing were unknown arts. The poor were clad in rags and skins—they devoured crusts, and gnawed bones. The day of Science dawned, and the luxuries of a century ago are the necessities of to-day. Men in the middle ranks of life have more of the conveniences and elegancies than the princes and kings of the theological times. But above and over all*

*this, is the development of mind. There is more of value in the brain of an average man of today—of a master-mechanic, of a chemist, of a naturalist, of an inventor, than there was in the brain of the world four hundred years ago. These blessings did not fall from the skies. These benefits did not drop from the outstretched hands of priests. They were not found in cathedrals or behind altars—neither were they searched for with holy candles. They were not discovered by the closed eyes of prayer, nor did they come in answer to superstitious supplication. They are the children of freedom, the gifts of reason, observation and experience—and for them all, man is indebted to man.*"[24]

We now see, though, that Ingersoll's claim is overly simplistic as the movement towards scientific thinking and science literacy was supported through the educational systems created for religious purposes. While Coyne argues that religion and science are incompatible, because there are two systems both aimed at understanding nature, we can see here that religious pursuits have the potential to actually support educational and scientific advancement.

Even though Coyne claims in the introduction to his book that his primary goal is not to prove that religion has only had a malign influence on society, the overall tone of his publication is antagonistic and divisive, epitomized by statements such as "above all, I will have achieved my aim if, when you hear someone described as a 'person of faith,' you see it as criticism rather than praise." Yet, again—how can we demand more flexible thinking from Bible literalists if we, ourselves, are not even able to concede that faith has its positive sides? It is therefore imperative that we as educators and scientists recognize and avoid the tendency to dehumanize, vilify or objectify creationists. Even if Dawkins and company find creationist *actions* appalling and their *views* ludicrous, we must still be able to see the creationist's humanity independent of the fundamental ideologies that they hold. While we can ardently disagree with their fundamentalist beliefs, we will have to learn to decouple the ideology from the person if we are to have any real chance of engaging with these individuals. For that reason, we will now turn to the personal stories from former creationists in order to better understand the complexity of this issue at an individual level. I humbly hope that these personal accounts will help resolve the trance of the unreal other.

## Beyond the Perception of the Unreal Other

*"If we could read the secret history of our enemies, we should find in each man's life sorrow and suffering enough to disarm all hostility."*

Henry Wadsworth Longfellow, The Complete Works of Henry Wadsworth Longfellow

Xenophobia is characterized by mistrust and aggression towards those who are perceived as different and is often the source and cause of stereotyping, prejudice,

---

[24]    Robert Green Ingersoll, "God in The Constitution".

racism, discrimination and general division in society. Yet, this discrimination goes beyond citizenship and race and can be understood more generally as the fear of those who are perceived as being 'others'. If we look at the degree of division that permeates our society, we will see that it is not only between individuals of different genders, races or nationalities, but far more we see that ingroup | outgroup discrimination also occurs due to varying worldviews, political views, or religious beliefs. The 2016 presidential elections in the United States and the subsequent election of President Donald Trump highlighted this tribal-like division within modern society and our quick discernment of who is 'us' and who is 'them'. News broadcasts and social media posts give us the impression that American society has become irreconcilably divided on the basis of beliefs, opinions, value systems and worldviews.

As just discussed, this same societal division is reflected in comments made by some biologists such as Jerry Coyne who claims that religion is inherently detrimental, dangerous and entirely incompatible with science (Tayler 2015). The creationists are undoubtedly also engaged in this same evolutionarily-driven 'us versus them' thinking, carrying around their own black-or-white belief statements such as: "Scientists are against God and religion" or "Teachers and scientists want to turn everyone into atheists" or "Scientists make up a bunch of 'facts' to achieve their atheistic goals." With a set of such convictions, it is easy to see how they come to the conclusion that they need to resist learning about evolution at every cost, especially when their fears appear to be verified by statements made by Coyne. The creationism | evolution conflict has caused parties on both sides to focus on our differences and forget all of the positive human qualities that those on the other side possess. Clinical psychologist Tara Brach refers to this phenomenon as the 'trance of the unreal other':

> *"The more different someone seems from us, the more unreal they may feel to us. We can too easily ignore or dismiss people when they are of a different race or religion, when they come from a different socio-economic 'class'. Assessing them as either superior or inferior, better or worse, important or unimportant, we distance ourselves. Fixating on appearances—their looks, behavior, ways of speaking—we peg them as certain types. They are HIV positive or an alcoholic, a leftist or fundamentalist… . Whether extreme or subtle, typing others makes the real human invisible to our eyes and closes our heart. Once someone is an unreal other, we lose sight of how they hurt. Because we don't experience them as feeling beings, we not only ignore them, we can inflict pain on them without compunction… . All the enormous suffering of violence and war comes from our basic failure to see that others are real."*[25]

So how do we change this? As much as we have tried and wished that the creationists would start—that they would wake up one morning and realize that science is right—this is not likely to occur as long as these apparent battle-lines are drawn. There must first be a way to have a conversation where both parties feel heard.

---

[25]   https://www.tarabrach.com/trance-of-unreal-other/Accessed 21 July 2018.

In my opinion, this is the point that must be reached between scientists and creationists. There must be a space made, where a true conversation is possible. Although I will never become convinced that religious ideology belongs in the science classroom, I am more than willing to see the creationist as a person who holds their beliefs for reasons that *they* consider credible and good. Their fear of science is real for *them* and we must recognize and acknowledge the reasons for that fear so that we can engage with them in a manner that does not confirm these fears but instead mitigates them. Effective communication will require us to depart from interactions that illicit a fight-or-flight response in a concerted effort to promote tend-and-befriend-based interactions. In an attempt to support this type of open and effective communication, we will now hear from former creationists who have been courageous and kind enough to share their stories, so that we can clearly recognize the humanity of the creationist and dispel the prejudicial claims that have been levied against 'them'.

## Conversations on Creation

I have spoken often of the need to differentiate between an ideology and a person—meaning that we recognize the creationist first as a human and not as a concept. While I also lumped *creationism* and *creationists* into the same category for many years, a personal conversation that I had with Pete Kron, my cousin-in-law, opened my eyes to the complex emotional nature of this belief system. Although he probably did not know it at the time, the brief conversations that we had were a real eye-opener for me. Hearing his own struggle to come to terms with his doubts awoke a great deal of empathy and understanding within me. I had read all of the books from and about creationists. I had watched their movies and signed up for their newsletters. I had tried to truly understand the creationist phenomenon but, in the end, I was still limited to my view from outside. Pete gave me my first idea of what it really *felt* like to be a creationist and what it was like when one decided to leave the church.

His story and his ability to reflect on the entire process of belief and doubt had a profound effect on me. For that reason, I would like to offer the reader a brief insight into those conversations in the hope of awakening the same understanding and insights that they afforded me. Following Pete's story, I have also included excerpts from my conversations with Kayla Rau who was also kind enough to share her story. Pete and Kayla grew up on opposite coasts in worlds that were very different from the other. But what connects both of their stories is the centrality of the creationism in their lives and their ultimate departure from this belief system. Their personal accounts shed light on the humanity of the creationists and enabled me to break out of my own 'us versus them' mentality. I hope that their stories all also awake similar feelings of connection and empathy in the reader. I would like to express my sincerest gratitude to Pete and Kayla for sharing their stories and for offering me and now subsequently all readers a glimpse into the intricacies of creationist culture. May your open-mindedness inspire the same in all who read your stories.

## Conversations with Pete

(Elizabeth) So I know that you are living in Tennessee now, did you grow up there? Can you tell me a little bit about your upbringing?

(Pete) I was born in 1984 and grew up on the North Fork of Long Island, New York in a ridiculously picturesque neighborhood with lots of kids my age. My family was very stable and healthy. We attended the Presbyterian church at the end of my street. Again, very picturesque New England church built in the early 1800's. I never attended a non-Christian school. I went to a very small but excellent elementary school, was home schooled for junior high, attended a private (not so excellent) high school, and finally graduated from Belmont, a Christian university. My parents sacrificed a tremendous amount of time and money to keep me out of the public-school system. At one point my father switched offices so that he could drive us to our high school every day which was over an hour away. Evolution was one of the specific reasons for [keeping us out of public schools].

(Elizabeth) So, church was clearly a major component of your childhood, is that right?

(Pete) Church was an extremely important part of my life. That actually seems like an understatement. The church was inseparably woven into the fabric of my childhood and teenage years.

My mother directed the church choir for several years, she was frequently hired as an organist in multiple other churches, she was our Sunday school superintendent for ten years, founded and worked in multiple church charities, held a prayer group/ Bible study at our home for years, helped start our church's modern worship group (that I was forced to play drums in), (laughs) and wrote/played original Christian music at concerts up and down Long Island.

My father was highly respected and held many positions at our church. He was basically in charge of the church's finances, he was chairman of several Christian charities, he also played in the worship band, he listened to many Christian apologetics on tape in the car (including all of those hours to and from high school), and most importantly he taught my sister and I all of our morals and life lessons through the context of Christianity.

This is really just scratching the surface, but you get the idea. I was crawling around under the church pews since before I can remember. Almost all of my friends and activities were linked to the church in some way. To sum up, I was constantly surrounded by Christian culture and rhetoric for 23 years.

I should note that when I talk about Christianity, I'm referring to traditional, conservative Christianity. Nothing super aggressive like Westboro Baptist, but definitely 'literal' interpretation, anti-gay, pro-life, pro-abstinence, anti-sex ed., pro-prayer in school, extremely anti-evolution, climate change denying, etc. The key point here though, and I want to emphasize this, is that all of this was wrapped in up a super positive, picturesque, happy childhood. My parents loved us. They donated tons of money. They served their community. They surrounded us with a community of other loving families and friends. We had wonderful family traditions. This is what people feel they have to give up when you present them with evidence for evolution.

A literal interpretation [of the Bible] is a house of cards. This is why it's so fiercely protected. The Bible is a book from God! If you give yourself the authority to remove something uncomfortable here or there, then everything is up for grabs. You are giving yourself authority over God and the book loses all credibility. So, it's an all or nothing approach. Your community, your family, your marriage, your traditions, your morals all collapse if you remove one card, such as the Genesis story.

I've had many people ask me what's stopping me from living some kind of wild immoral life now that I don't have God to anchor me. They have received all of their moral teachings through the church and so they cannot separate the two. In Sunday school we used to joke that if you're not paying attention and someone asks you a question just answer, "because of Jesus". It's always the right answer! But in all seriousness, we now have generations of adults who are still answering "because of Jesus" to all of life's hard questions including Genesis. So, I'm not offended when I get that question. I just see a grown man or woman that hasn't had to do the philosophical work of justifying their values and morals. They are leaning on external validation rather than innate value. I like to put myself out there as an example of a nice, hardworking, boring guy who just doesn't believe in God (laughs). They usually don't know what to do with that. But this is really key. It doesn't matter how many fossils you show them. If they except something like evolution, their whole world will collapse. They're cornered. There needs to be a door, a path, a vision for them that feels safe.

(Elizabeth) Do you remember learning about evolution?

(Pete) My earliest memory of evolution: I think this was in second or third grade. It came up in class and my elementary school principal said that she was "not willing to discuss it" with any of us. The class immediately erupted because this meant that she believed it! I remember thinking, "Well if you're a Christian, there's nothing to discuss." AKA you can't be a Christian and believe in evolution. Like I said this is my first memory, so clearly all of us must have been indoctrinated shockingly early.

Later on, in 7th grade I was introduced to Mr. Ken Ham and his 'wonderful' ideas. This was really where I took my first in-depth look at evolution. We actually heard him speak in person. I thought he was friendly and funny. He made a lot of jokes about his accent and how he looks like Abraham Lincoln. And he gave us ammunition to fight off uncomfortable ideas that we felt literally threatened our way of life.

Finally, in high school, New York state required that I receive an actual education about evolution. So, our science teacher taught us what we needed to know to pass the New York state exam, and then devoted a full two weeks to teaching why it was all false. And this was very likely the best teacher at that school.

I remember one of my friends was smart enough to quietly question what we were being taught. But he had an unfair advantage. His father was not a conservative Christian. And his mom and stepdad were not living the values that they preached. So, for him the cultural bubble had already popped to some extent. I was still fully immersed. So, while I respected his critical thinking, I still couldn't let myself go there.

While I was 'immersed' in Christian schools and cultures, there were other kids who took their beliefs to public school. I knew one guy from our youth group who intentionally failed biology. Believing himself to be a martyr, he refused to correctly answer any questions related to evolution. His parents praised him. I could tell other adults weren't so sure. But no one spoke up and said, "This is crazy!"

(Elizabeth) What caused you to begin to question what you had been learning?

(Pete) I think I began to question in college or maybe even senior year of high school. I'm not sure I could really pinpoint a specific moment. But the first time I voiced my doubts as by accident. Our professor required us to break down texts and highlight classical rhetorical elements. In one of my examples, I dissected conservative, religious rhetoric. In particular, I pointed out how an emphasis on Ethos via divine authority allows for cyclical reasoning and twisted logic (Logos), i.e., using the Bible to prove the Bible. I was surprised at what I had written because I still considered myself a Christian, and a conservative one at that. But clearly doubts were beginning to surface.

When my teacher returned my paper, I noticed that it had a note at the bottom. It said, "This is exactly how I feel! But you need to be careful with that stuff around here." So yet again I discovered a highly intelligent and qualified educator closeted by a conservative environment. One can only wonder what happened to her that she felt the need to issue such a warning.

I believe one of the biggest problems that closeted doubters face is isolation. When I was with my Christian friends I would hint at my doubts. They would hint back about their doubts to sound open minded. Sometimes someone would quote something like 1 Corinthians 13:12 "For now we see through a glass darkly; but then face to face: now I know in part; but then shall I know even as also I am known." Generally, there was a lot of talk about "embracing the mystery" which feels gratifyingly deep and avoids the obvious questions. But in my experience, no was ever candid enough to say, "Hey this isn't a mystery, it's a contradiction!"

On the other hand, I definitely couldn't talk about my doubts to my non-Christian friends because I was supposed to be 'witnessing' to them. Remember a very literal Hell was on the line! Even though I wasn't sure I believed in it, I wasn't going to risk showing any cracks in my faith. This is just one example of how the conservative bubble is preserved outside of religious environments. Isolation forces a doubter to process everything internally. It was just me and that voice in the back of my head saying, "this doesn't make sense." But that voice can easily be suppressed by social pressures, fear of the unknown, an emotional life event, or even just a busy lifestyle.

I remember a couple of my good friends were having a classic talk about "embracing the mystery." I was so tired of it that I called them out. Things got interesting for a minute. Then one of them suspiciously asked, "Pete are you still a Christian?" I wasn't emotionally ready to "deny Christ" (which believe me feels every bit as dramatic as it sounds). So, I didn't. Also, they would have instantly tightened up with an outsider in the mix. I quieted down and just listened for a while. I remember thinking, "This same conversation is happening somewhere in another country, with the same questions, the same personalities but a different religion." I think that was a big moment for me.

But for every logical break through there was another emotional snare to untangle. I should mention the issue of loyalty. I definitely remember feeling guilty.

There was a comradery that came with 'keeping the faith', standing strong together when others criticized our beliefs. It's hard to deny the evidence and 'embrace the mysteries'. And when a man was lost, it weakened our numbers. I believe there are many doubters out there sick to their stomachs with the prospect of betraying their friends and family.

(Elizabeth) When did you finally decide to voice your doubts to your friends and family and how did they respond?

(Pete) This may be too personal, but I found an old group email where I'm coming out as a nonbeliever to my college buddies. I think it sums up what I was feeling at the time:

> Friend 1—"I was thinking about you guys today and thought we should keep in touch with a weekly, biweekly e-mail, nothing long or crazy just kinda like an update on what we are reading, listening to, and maybe how we can encourage each other and or prayer requests."
> (… He goes on to list music, books, prayer requests. We all did the same. And then I went on to explain…)

> Me—"I should probably also mention as far as prayer requests, I'm not a Christian anymore. It's tough to say that to my Christian friends. It really feels like a breakup or a betrayal of some sort, but the reality is I'm the same guy. I feel no bitterness or desire to lead some crazy life. I just came to grips with the fact that I didn't believe and finally gathered the courage to walk away. In fact, it's probably the most courageous, honest thing I've ever done in my life. I have a lot of family and peer pressure that makes me want to keep quiet but I'm trying to be more open. So there ya have it."

> Friend 1—"I'm glad you were open enough to share with us about your decision. Would you consider yourself an Atheist? What were some of the things that led you to finally walk away? I'm sure you already know that while I'm disappointed to hear you don't believe, I still love you greatly. It's crazy because we haven't talked about God in such a long time. I want to know more about how you came to the place you're at now."

> Friend 2—"Pete, I appreciate the honesty that you put forth in your email below. I have not been a friend to you by not knowing or seeking to know about this area of your life. I am very sorry. I would really love to see you when I am in Nashville for the 4th of July. I want to catch up, and I want to acknowledge that I would like the chance to speak to you specifically about Jesus, faith, and the Church. I know that sounds awkward, but let me know what you think."

> Me—"Friend 2 I'd love to hang out when you're in town. We can talk about whatever you want. I'd just be excited to see ya.

> Friend 1—I'll try to answer your question about religion. I guess it's kind of like being a Mets fan haha! Ya know when you're a kid you go to games and you love it: you've got the hat, your family is there, you meet friends, you really

believe they're gonna win every game, the grass is green, the logo is beautiful, there's a romance about it, a culture, a tradition and most of all an identity. "Mets or Yankees?" "Oh I'm a Mets guy for sure." Anyway, you get older, understand the game more and more, the players, the coaches, the other teams. By the time I was a senior in high school it was the subway series. Huge. I remember hearing someone explain how the Mets were statistically superior to the Yankees and while I was listening, I knew in my heart that the Mets were going to lose. But of course I rooted harder and more belligerently than ever because I didn't want to doubt. But I knew. Before game one, I knew. To this day I know maybe one player on the team and yet I have a Mets debit card. Why?! It's an identity. I'm a Mets guy. Anyway, that's probably a ridiculous analogy but kind of funny. It does sum up the feeling of emotional attachment and identity. Also, the need to fight to believe even when in the back of your mind you doubt. It's especially urgent when there is an opponent as villainized as the Yankees or even Atheists. I consider the word Atheist self-defeating. A-theism means not-theism. If you are a theist, it says not you. What a futile way to start a discussion! 'All your core beliefs, your morality, your community, I'm not-that.' So, it's a divisive, inaccurate word. It is also associated with a-moral and carries innate negative value. Furthermore, it establishes Theism as the root which is my main beef, which I'll explain. So Atheism says virtually nothing about what I DO believe and misrepresents what I DON'T believe. I don't know if I've come up with a word or category or if I have to. I'd rather just be Pete. That should say enough. But I've thought about the word Naturalism. Steinbeck called it "the Is". To seek out and make peace with truth regardless of how ugly or beautiful it turns out to be. I want to start with what we can know. If people want to add to that, in other words tack on anything Super-Natural, that's great and I'm open to it. But the burden of proof is on them. Even if I come up with a good theory that I'd like to live by; awesome but, I could never assert any imperatives from a theory or something faith based. The problem I have is that we start with super-natural claims and work backwards. Just like I was a Mets fan before I knew what baseball was. Even non-religious people are usually raised to respect other's religious beliefs. The credibility is grandfathered in long before we learn how to think critically. And in my case, emotion and identity cemented a commitment to super-natural beliefs for years after I thought to question things like angels or what have you. So to sum up. I stepped out not knowing what I'd find, but there are logical, knowable, natural paths to many of the same morals and community-based values that I believed as a Christian. So that's what I mean when I say I feel like I'm the same guy, just different paths to the same conclusions. I feel like I stepped into the unknown and turns out everything is still there. I suppose I did have to give up the afterlife. But the possibility of nothing happening isn't as scary as the probability of billions in Hell. So I don't feel a loss there. I feel aggressively optimistic and fully responsible for my decisions and fully grateful for those around me.

Anyway, I didn't want to go on for that long. And I certainly didn't want to write an entire email about religion. But it's a very broad, interesting topic and my views have evolved dramatically since we've really hung out. There's just a lot to clarify. If we were all originally non-believers and someone became a Christian, we'd want to know if they were liberal, southern baptist, catholic, etc. I appreciate you guys wanting to know and I'm totally open to talking about it if anyone has any questions. Mostly though I'm just stoked to reconnect."

And this is where the conversation ended. There were a couple more emails, but no one ever responded or wrote anything about religion. 'Friend 2' never contacted me when he was in town. My lengthy reply established me as an outsider, not someone on the fence. I've experienced this many times since. Rather than a healthy discussion, we shut down. They chalked me up as a lost cause and I didn't want to offend them or distance myself any more than I already had. That's it in a nutshell. Whether it be religion, sports, politics, etc., most of us don't want to either engage with or be the outsider.

Concerning family, they were obviously the hardest to disappoint. In fact, that was the first word my sister and brother in law said when I told them, 'disappointed'. I desperately needed to tell someone in my family. Participating in religious traditions such as prayers before meals and special church services made me feel like a liar. So, I blurted it out one day like an awkward breakup. We were talking about how I'd been 'church hopping'. It's a common thing in Nashville. There are just so many churches. Another reason is that people aren't comfortable with their faith. Nashville is an increasingly liberal town with conservative traditions. So, people hop from church to church hoping to find a compromise. Given my background though, I couldn't bend so I broke. I told them, "I've realized it's not the churches I've had a problem with this whole time, it's the religion!" It just needed to come out and it didn't matter how! They expressed their disappointment but let me know that they still loved me. This was a great reaction! I truly cannot complain given what many people go through. However, it was strange being, in a way, forgiven for my deepest beliefs. It was odd feeling that this reaction was a generous response. There is truly an authority that comes with being in the majority. That authority proved to be very thin though. The tone shifted as it became clear that I was confident in my non-belief. Once again, the conversation ended quickly, and we've never talked about religion since.

I did briefly try to talk to my sister about evolution. I was as kind as possible, knowing that it would be a sensitive topic. But she teared up immediately and told me, "Evolution just really doesn't affect my life." In other words, "I don't want to talk about it!" Since then they've decided to homeschool my niece and nephew. My niece has turned into a little biologist. She can't stop herself from picking up every slimy bug she finds. She is great with animals and is just a sponge for any information she can get about nature. So suddenly evolution is very relevant in my sister's life. And I regret to report that they have already been to Kentucky to see Ken Ham's life size Noah's ark. It's easy to feel disappointed, just as they were with me. I would love to beat them over the head with logic and rattle off 20 examples of how idiotic Ken Ham's theories are without taking a breath. But I have to remember that evolution is a direct threat to their traditions, to peace in their family, to a moral path for their children, and frankly to their marriage. I don't know if I could have walked away if I had so much skin in the game. If the Bible is God's literal word, then there is no room for compromise. Ken Ham just happens to be the most visible, vocal and confident champion for their values. Logic and fossils really have very little to do with it. "Were you there?" is the perfect quick fix people a looking for.

Part of the reason I came out to my sister and brother-in-law was because I hoped that word would spread to our parents. This seems so weak in hindsight. I should've been able to be direct with them. It's important to understand that in conservative religious cultures, raising your children to follow God means the world. In fact, it means this world and the next world! I remember when my best friend asked his parents if they could be happy for him if he was a healthy, successful Buddhist. Of course, they said no. How could they be happy knowing that their son was "worshiping a false god" and was bound for eternal punishment. The family suffered tremendous conflict for years; the kind of conflict I was hoping to avoid.

I did end up telling my dad over the phone one Easter Sunday. He asked where I had gone to church. I told him, "Dad I'm just really not religious anymore." He paused for a moment and said, "Well Peter, I find that I'm a much better person when I go." When we got off the phone it occurred to me that he thought that I had been sleeping in or something. In his mind, I hadn't rejected the faith! I was just lazy! Someday I'd get serious and come back to God. Basically, he autocorrected what I said to something more palatable. I had a similar experience with my mother years later. The phrase "flag burning Atheists" may have been used. I laughed and said, "Mom you know your son is an Atheist, right?" She corrected me and said, "well Agnostic" then quickly changed the subject. I had never once mentioned Agnosticism! But she had apparently made peace with the idea of me being in a time of searching. So, for years I lived as the religiously irresponsible/questioning son. This was better than what I saw as hopeless arguing during our brief family get togethers.

Very few people are looking to change their belief system in their fifties and sixties. And that's nothing short of what I would have been asking my parents to do. There is no arguing with the word of God. John 14:6 "Jesus saith unto him, I am the way, the truth, and the life: no man cometh unto the Father, but by me." This is core doctrine. It does not allow for your child to get into heaven by being anything other than a follower of Christ. Period. I wish I could say that I've found a bridge over these kinds of scriptures, but I haven't.

Please understand, my family, as well as my friends' families that I have mentioned, are highly educated, intelligent, loving, and generous people. They do not resemble any strict, angry religious stereotype. We need a better path for these families who don't want to believe that their loved ones are going to Hell. I did finally have some candid conversations with my father. Before he passed away, we had quite a bit of downtime together in the hospital. He's the one who finally brought up the subject of religion. I wasn't particularly motivated to start arguing against his belief system while he was laying there on his death bed, and I told him as much. But there he was immobilized, pale, dying saying, "Come on Pete don't hold back, I can handle it!" He said, "Pete I might surprise you. I want to know the truth. I don't think most people do. I don't think most people can't handle the truth. But I want to know, good or bad." It was really the perfect opening for me. I told him, "That's the exact the thought that freed me to leave the church." I remember listening to the pastor scold the congregation about how we need to work to find reasons to believe the Bible not the opposite. This went against all I'd ever been taught about critical thinking and analyzing texts. You don't start with what you want to believe and work backwards. I thought, "If God is truth. And all truth comes from Him. Then if I'm not pursuing the truth by all of the means that I have available, then I'm not pursuing God. In fact I'm sinning against Him."

My dad and I talked quite a bit. He had me listen to one of his favorite Christian apologists again (captive audience). We went back and forth a little bit. It mostly went nowhere and was somewhat frustrating. We ultimately moved on and just enjoyed each other's company. But I later thought of something he used to say all the time: "We are the books we read and the friends we keep." He was famous for his large lending library of self-help books and wisdom literature. In hind sight these bookshelves were a timeline or a map of his search for truth, as many of our bookshelves are. My father's search was very deep but narrow. Almost all of his books were on either business or Christianity. And that was him. They served him well. But if there is a solution or a bridge between modes of thought, if there is going to be constructive communication between any differing opinions we're going to have to tend to our bookshelves. If a person is really after the truth, then a wider search can only strengthen their position.

(Elizabeth) How has your experience affected the way you see religion and science now?

(Pete) So many of us live in a bubble. We only let down our guard with people who believe what we believe. There is safety in numbers. But the feeling of power and authority that comes from being in the majority is dangerous. This condescension is identical in religious and scientific communities. It prevents us from having a real conversation. I am a fairly quiet person. It's amazing what people will tell you when you let them do all the talking. When I was a conservative Christian, I had Atheists rant to me about boring, idiotic Christianity. Conservatives were always 'those people'. Now that I'm not religious, people assume I'm a Christian all the time. One guy I worked with said he knew I was 'born again' because he could see Jesus in my eyes. Republicans think you're a Republican. Democrats think you're a Democrat. It's all because no one can fathom that there's a friendly, reasonable person on the other side. It's really easy for us to demonize groups of people that we've never met or believe we've never met.

If no good ever came from religion, it wouldn't exist. There's a reason why people are drawn to it. Obviously being cleaved from the herd was not fun. I have many more negative experiences that I could talk about. But I need to say that there are good reasons to be religious. The community at its best is a wholesome, positive, family orientated environment. Many churches are full of intelligent, moral, generous, and loving people actively working on themselves and serving their communities. The church offers a structure for life. For every situation, good or bad, there is an answer and a support system set up and ready to go. An organized and caring community is a powerful thing.

I also believe that a spiritual practice is important. I'm reminded of the Carl Sandburg quote, "Time is the coin of your life. It is the only coin you have, and only you can determine how it will be spent. Be careful lest you let other people spend it for you." We encounter dozens of people every day and maybe thousands of advertisements all grabbing at us, trying to spend our time and influence us. Setting time aside weekly or even daily for a spiritual practice is the greatest defense that I've found against these forces. However, it has been difficult to find this kind of consistent practice outside of a spiritual community. There is power and accountability in numbers. It has been a struggle to go it alone.

Today I'd still consider myself a Naturalist. Although I've developed somewhat of a crush on Buddhism! (laughs) I greatly admire Buddhist philosophy. This is in part because I have no personal connections to any fundamentalist communities. I feel free to interpret any supernatural beliefs, such as reincarnation, as useful metaphor. I suspect that this is similar to how many liberal Christians feel. I have no doubt that Christianity can be and is a beneficial practice for many people in this context.

I've lived on both sides of the fence. I've had many discussions with all kinds of people. What I've found is that most people know little to nothing about the other side. Furthermore, most are not interested in understanding the other side. So, I believe the first question really is, "do you want to know the truth?" If that is truly what's motivating the discussion, there should be a thirst for differing opinions. All Christians should have a well-worn copy of *On the Origin of Species*. All non-believers should interact with religious communities. There should be no talk of "those people". Like a chess player walking around the board to see his opponents view of the game, effective persuasion is rooted in a deep understanding of the audience. So, the most intense evangelist, religious or otherwise, has every reason to broaden both their reading and their friendships.

I still struggle with when and how to talk to people about religion. Unfortunately, it's still hard to talk to family and friends. But if someone like my niece is truly curious, I don't need to talk to her about Ken Ham. If and when my she comes to me asking questions, I'll just tell her to walk around the chess board, read as much as she can and seek the truth by all means available. If the truth is what she's after and truth is what my family wants for her, then there is nothing to be afraid of.

## Conversations with Kayla

Kayla and I met in Costa Rica at an animal rehabilitation center for exotic animals. In other words, both of us had spent hundreds of dollars to travel across the world to clean enclosures and refill water dishes and both of us loved it. During one of our breaks, we began to discuss what we did back home and my work on creationism led to our discussion of her creationist upbringing and her eventual departure from this belief system. Below is the transcript of our subsequent conversations that she has kindly agreed to let me reproduce in this publication:

(Elizabeth) Could you briefly describe your childhood—when and where did you grow up? What type of school did you attend?

(Kayla) I was born at a midwife's clinic in a small town in southern Oregon and was then whisked away to my family's hilltop home in a much smaller town where I lived for the next twenty years. We didn't have any neighbors—the nearest 'neighbor' was about half a mile away down the road. I was homeschooled along with my two older sisters and our younger brother. So, we were really isolated. I am pretty sure that my mom decided to homeschool us because she didn't want us exposed to evolution, at least that was one of the main reasons. She had been raised Catholic and then 'found' herself later in life when she joined a Christian church. So, she taught us at home and the curriculum was a Christian one. She even administered our yearly tests at home, which wasn't technically allowed but she fudged the information on where we had 'officially' taken the tests, saying that we had taken the exams at the local elementary school, but that wasn't true. So, we never went to the school (even though it was located just down the hill).

(Elizabeth) So, did you have any contact to other kids?

(Kayla) I did—at church. Church was my only socialization. That, and my sisters and I did take tap/ballet dance classes from when I was 6 up to age 12, but that wasn't really socialization. The only regular socialization I had throughout my childhood were activities connected to the country church, which my mom started taking us to when I was three—Sunday school, junior church during the Sunday night service or a Wednesday program called 'AWANA', which stood for 'Approved Workmen Are Not Ashamed'. So that is basically where I got all of my socialization. I mean we also went to this roller-skating event for homeschooled children but it was the worst because we were all homeschooled and all painfully shy and therefore spent the entire time trying to avoid eye contact.

I started attending community college at age 16 and was taking full-time classes by 17, and it was there that I had my first real exposure to secular lifestyles and people who weren't all Christian.

(Elizabeth) So this was your first time exposed to secular lifestyles. What was that like?

(Kayla) Yeah, my first real exposure. I mean long enough to realize these are all fully functioning, well-adapted people. They are not bad people (which was counter to what I had been taught). I had never been around it long enough before to start making those connections. Well, honestly, I didn't even make those connections until later. At the time I thought they were a bit scary and something was wrong with them. But I was still a 'kid' and I didn't know how to think for myself yet. [Side note: I didn't hear the word 'fundamentalist' until I was around 18, when a friend at the community college used the word and I was like 'What is that?" and he looked at me and politely said, 'That's what you are!' (laughs)].

So, I lived at home during this time, until I was twenty. For college science classes, I only took chemistry and earth science. Chemistry did not make much of an impression on me, but earth science was so enjoyable (because of the teacher) that I took an additional two terms that I did not need to complete my AA. Considering that rocks were the major focus of the class, we regularly dealt in non-biblical measures of the earth's age, but I don't recall ever challenging the teacher on that or any other controversial subject. I spoke up often in class, but only ever to give an answer I knew was right or ask a question for clarification. I was too intimidated by persons of authority to challenge them, even when they flew in the face of what I'd been taught. I'm sure I corrected the teacher's words in my head, but I sat in silence, content that it was enough for me to know the 'right answers'. This seems contradictory to what Christians are taught (e.g., to spread the word of God and Jesus and salvation and everything else in the Bible). All I can say is that, clearly, the idea of an all-seeing, all-knowing God coupled with 'Obey thy father and mother' and other such submission tactics made more of an impression on me than the command to proselytize. Then when I finished three years at community college, I transferred to Western Oregon University, where I lived near my oldest sister, and I began to attend her church.

(Elizabeth) How important was church? Religion? God? Bible? For you and your family growing up? What role did it play in your life?

(Kayla) Church was extremely important to me, because it was the only world I knew thoroughly outside of my family's property. I mean my mom started taking us to church when I was very little and all of my formative years were spent in church. Church is where I got all of my firsthand information. There, I was well-liked by the adults because I was intelligent, polite, and outgoing with them. That gave me a good source of self-confidence, which was really crucial because we didn't have a lot of opportunity to build self-confidence being homeschooled. But this was also counteracted because church was also my main source of socialization with my peers, and I was really awkward with them because I was too smart. So, it was a counterbalance, all of the adults liked me just fine and thought that I was a good person but nobody my age wanted to hang out with me. So that was hard.

So, I found some affirmation at church, but I didn't come out of it as a self-confident, well-adjusted person—at all. I definitely learned all of the principles by which I lived my life there. And a huge justification for the way I lived my life as I did. [Side note: (Elizabeth) What kind of justification? (Kayla) Oh, to judge people. To know that 'I am doing it better than you' because I am doing things the way that the bible tells us and the way Jesus instructed us, so I know I am doing it the *right* way. I was capable of memorizing a vast array of Bible verses and trivia, which the people in charge liked. So, from religion, I drew the principles by which I lived my life, and I also found a vast amount of justification for living as I did, which is deeply satisfying in a circular kind of way.]

I was baptized when I was twelve. It is not like a Catholic baptism where you are a baby, but it is like a public commitment to live as a Christian. We'd have a little church service down by the Applegate River and then anyone who felt like they were ready to be baptized would be baptized in the river.

So, in general, church was very important to mom and me and my siblings, especially since that was the source of all of our socialization. Plus, it was really easy to integrate it into our life because it supported what we were learning at home in our Christian curriculum—you know very circular thinking. Everything supported this *one* way of thinking and this *one* particular lifestyle. It was literally the only thing we knew.

(Elizabeth) Why do you think that it was so important to your parents that you were raised in this way?

(Kayla) Church was important to my mom, not so much my dad. She and my father were both recovering Catholics when they met. But my dad was also a marijuana user and grower, and my mother needed saving from that, I suppose. I believe my mom found herself in the Protestant church. But I think the main thing was community. The life of a marijuana grower is very isolating, she lost a lot of friends, plus she has always been very shy. I think the church, God and Jesus helped alleviate her loneliness and frustrations. She met a lot of very kind people there. I think my dad played along to appease my mom, and because we kids liked going. He even sat on the board of elders for several years at one point, but I confess I've never had the nerve to ask him what motivated that choice.

(Elizabeth) Did personal prayer play an important role in your life?

(Kayla) A very big role! I prayed morning, every night and all throughout the day. [Side note: That was honestly one of the hardest habits to break when I left the church. And one of the things that actually led to some of my doubts about the church in my twenties when I realized, 'I have no proof that I am actually talking to someone! I might have been talking to myself for the past twenty years!' And that was humiliating and embarrassing in a very private way when I realized that this might have all been a figment of my imagination—this might all be fake! So, I don't pray anymore. For so many years it was presented to me as the solution and then I wasn't seeing any results. So, I why would I put any more energy into that when it's not providing any turn around? I now rather put my energy into what will really help these situations. Like when someone says, 'My prayers are with so-and-so' and I think—well maybe they'd prefer a casserole.] So, there was a lot of prayer! Plus, there was a lot 'taking prayer requests' and that became a lot more as I got older, especially when I was in 'youth group' and 'college group'. We'd sit in a circle and talk about things that were coming up, or about sick family members and then someone would take that prayer request. And we just went around the circle talking about where we needed help in our lives and then we would all pray for each other.

So, I prayed to God and Jesus every day, and I was constantly motivated by what God would think of what I was doing, how I was behaving. Religion, God, and the Bible were all very important to me, but I'm still trying to figure out exactly why, and why those priorities persisted for so many years of my life.

(Elizabeth) What is your earliest memory of learning about creation or about the age of the Earth? Do you remember how you felt about this?

(Kayla) I don't remember *learning* about creation, because it was just always there. But I do remember learning about 'Day and Night, Sky and Water', etc., when I was in the 3-and-4-year-olds' Sunday School. There was a little chant we were taught with the order of creation as taught in Genesis, and I can still remember it now, "Day and night, Sky and water, Dry land and plants, Sun, moon, stars, Birds and fish,

Animals and man, Then God rested". The whole purpose was to teach really little kids the creation story in Genesis.

(Elizabeth) So, Genesis was an important part of early education?

(Kayla) Oh, yeah, especially about the part where humans were the culmination of creation. Which is a little embarrassing now. (laughs) I suppose I was too little to have any reaction to this explanation of the beginning of everything. Add to that the fact that I was extremely trusting of all adults as being reliable sources of information—because they were *Christians*—and so I swallowed it as pure fact from day one. I must have been in third grade or so when I first heard that the earth was supposedly 6,000 years old. I don't have a clear memory of when it happened, but I seem to remember feeling in awe that somebody could name the age so confidently. To a grade-schooler, 6,000 years also seems like infinity, so I'm sure that blew me away, as well. Of course, it turned out that they probably *didn't* know and that they were just guessing based on genealogies and second chronicles and such.

(Elizabeth) Do you remember around when and where you first heard about evolution or about biological origins of humans? Do you remember how it was presented to you?

(Kayla) I don't remember exactly when or where I first heard about evolution, but I'm sure it was presented in a negative light. Laughable, even, like they would present some idea of evolution and then follow it up with the statement, 'But we all know that that's not true.' But I think it must have come from either my textbooks or church, because neither of my parents likes to voluntarily discuss anything that contradicts their own points of view. They would rather never bring up such topics.

(Elizabeth) At that point in your life—how did you feel about evolution? Did you notice that it conflicted with what you had learned about creation? Was it easy or difficult to assimilate scientific and religious ideas about human origins?

(Kayla) At that point in my life I considered evolution ridiculous, as I was taught to do. I knew it conflicted with the creation story, ergo it had to be false. There was no assimilation at all, only dismissal. This was very easy to accomplish, sheltered as I was from all people who believed theories other than creationism. There was, quite literally, no on in my life to argue for any idea that contradicted what I was learning from the Bible. But in general, I didn't think about it, because I was taught not to think about it—for like eighteen years.

(Elizabeth) What caused you to begin to question the literal interpretation of the Bible or what you had been taught in the church?

(Kayla) It was a slow road—it took years, maybe eight years. Once I became a full-time student at my community college, I made a few friends who were not Christians and who were very good to me. I was fortunate enough to have a repeat of this process at the university I transferred to. At that time, I was regularly attending a new church for the first time in my life, and all my friends there were new, too. I was majoring in English and minoring in theatre, and this latter interest caused a series of little rifts between me and my new Sunday School acquaintances especially during prayer requests. So, any time an audition or tech day was approaching, I would say that I was nervous about it and ask for prayer for the event, or I would say that it was a tech day and I that I wouldn't be able to go to church that day, and I started noticing that any time I talked about theater, the room would go silent and that nobody was interested in accepting the prayer request. I realized at some point that it was intentional and that they were ostracizing me. People who were otherwise sociable with me would turn quiet and distracted. I also noticed a silent disapproval, and I started feeling frustrated about all of that. I wasn't doing anything wrong by acting, yet people were treating me as if I was. With all my years of training in biblical mores, I knew that I wasn't doing anything wrong because I couldn't think of a single statement that would condemn Christians for taking an interest in theatre. There are, however, several passages that condemn tattoos, short hair on women, revealing clothing, etc.—things that are part of normal life for counter-culture folk, just like theatre and other art. Fundamentalists cling to these strictures but I felt that my churchmates, fundamentalists as they were, were taking rules too far and judging me for something they couldn't prove was wrong. I'd already experienced years of that exact same judgement from my mother, and I was sick of it, so I started feeling pretty disgusted once I figured out what was going on. I'd never had high self-esteem, and it upset me so badly that people who called themselves my friends seemed set on bringing me down about something that was truly inspiring to me. At the same time, I started noticing how supportive my non-Christian friends were of all my interests, how they were much more accepting of me. They were kind to me and they didn't ostracize me, even when I talked about church. It made me start to wonder if maybe they'd figured something out that I and my fellow Christians hadn't.

So, after months of the discrepancies between my two camps of friends gnawing at me, I drove to my church one day to knock on the pastor's door. I'd never visited his office before, so it would normally have been an intimidating circumstance, but he was a very nice man, and I was irritated enough that I wasn't afraid of anything at that moment. This was our conversation.

Me: I have a question for you, and I need an answer.

Pastor: Okay.

Me: Why is it that my non-Christian friends are so much more accepting of me than all my Christian friends? I'm different from them, but I'm not doing anything wrong, and yet I feel like I get so much more love from all my friends who don't follow Jesus. Why is that?

Pastor: I don't have an answer for that. I'm sorry. I wish I did.

Me: That's what I thought. Thanks for your time.

And I walked out. Thank goodness, the man gave me a straight answer and honestly seemed ashamed that there was no truthful defense to be argued. I think he knew that is the way Christians are a lot of the time. And there is no justification for that. Following the path of Jesus should be about love and that is what it came down to. 'Why do I feel love from this group of friends [non-Christians] and not from this group [church friends]?' It was a big moment for me. I lost faith in my fellow fundamentalists that day. It still took another few years after that for me to be open to other origin stories, though.

I think it was the next winter that I went on a mission trip to Mozambique to teach conversational English for a couple weeks. While there, I attended a couple church services. They were as foreign as a tribal language to me: the songs went on for countless repeats of verses and choruses, and *everyone* could sing. The sermon, as I recall, was about two hours long, too. And everybody was completely sincere and satisfied at the end of the whole ordeal. This caused further doubts for me. I thought to myself, 'These people practice the same faith I claim in a completely different way, and yet they're obviously finding total satisfaction in their methods. So, who am I to say that the way I practice Christianity is the only right way?' And then it was just like an alien thought in the back of my mind, 'Well, who am I to say that the way I do anything is the only right way?' I was twenty-two at the time and realized truly for the first time that people can live happy, fully functioning lives while following lifestyles very different than my own. Now, a decade later, I'm so relieved that a long-dormant synapse of justice fired just then.

(Elizabeth) So, your doubts didn't come about because you were exposed to evolution?

(Kayla) No, it was about the people. It started with my church friends and progressed when I left the country a year after graduating to teach English again. I went to South Korea, and the Koreans in charge of the school I worked at were literally the cruelest people I had ever met—and Christians. After months of enduring their abuse, I got really jaded. I really started questioning the validity of Christianity, given that my bosses and coworkers all called themselves such but were so mean-spirited and went out of their way to make my life miserable. I began to doubt a number of fundamentalist teaching. I dismissed the idea that homosexuality is a sin—I figured, if you ever find love, you better hang onto it no matter who it's with. I was so exhausted from my job that I stopped doing anything with my hair, which had hung down to my butt for my whole life. It looked terrible, all the time, and I got so sick of it that I stopped caring about the biblical commandment that women should have long hair and whack—I cut it all off in one go. I started drinking alcohol now and then, I started trying out swear words— I even learned how to lie, out of a necessity to protect myself from the people I worked for.

But the final nail in the coffin came when someone back home whom I cared for very, very much passed away. We had been in a relationship while I was at Western Oregon University, and he died while I was a million miles away. He was a loosely practicing Buddhist, and I suddenly found that I could not accept the idea that he was now 'burning in Hell.' I literally couldn't say that was a thing[26] anymore. When I finally came home, I was like, 'I just don't believe any of this anymore'. None of

---

[26] 'Thing' here is in reference to the rules that require a person to have accepted Jesus Christ as their Savior in order to get into Heaven.

this can be real. For so many years, we had focused on the afterlife and salvation and 'meeting your maker', and I just couldn't accept that version of what follows death anymore. I knew how much I loved him and how much he had loved me and what a wonderful, incredible person he'd been, and the idea of him 'burning in Hell' was such an injustice that I refused to believe it. There was no backtracking after that.

(Elizabeth) What type of emotional and social consequences did your doubts have for you? In other words, did it affect your general faith? If so, how was this for you? And how did your family and friends react to your doubts?

(Kayla) Yes, my doubts definitely affected my faith overall. After being a fundamentalist for so long, forcing the pieces to fit, the structure visibly started to crumble after that first stubborn piece refused to be wedged into place. I was attending a public university when I started doubting, so in some ways, the transition was easy because my new path was all around me. At the same time, I had been religious in the same manner for twenty years, and I really hated being told I was wrong, so I fought changing when challenged about it outright. This caused problems in the relationship I mentioned earlier, because he was eager for me to open up to new ideas, and I just wasn't ready to cave so soon.

When I finally broke fully from the church, I had entered a new realm of 'sick-to-death of the hypocrisy', so it was easier to be apart than a part of it. My family wasn't aware of my misgivings. By the time I was fully estranged from the church, I had broken contact with my sibling for other reasons, so they couldn't say much. When my mom figured it out, she was disappointed and hurt, which I have always found weird, because she'd had a similar transition from Catholicism to non-religious to Christian at the same age as I was. It took her years, but she has gradually become more accepting of me and my lifestyle, like that I'm living with a man I am not married to. (laughs) The nice thing is that since I have become a non-conservative, she has also started branching out and has even become friends with a gay man, which is such a big thing because homosexuality was such a 'horrible thing' for us growing up. I am glad to think that I have been some of the impetus for that since I am so different and have not followed the path that she wanted for me. It did break her heart, but I was sick of it. My father was supportive of pretty much all my new habits since he'd pretty much stopped going to church by that time anyway. My relationship and faith in God basically ended abruptly as soon as I left the church.

About five years after I left the church, I visited a prehistoric exhibit park featuring dinosaur statues, huge ferns, etc. I went for the dinosaurs, but what got me the most were the giant signs that walked readers through the Cretaceous and other eras, everything pertaining to pre-human existence. It was probably the first time since I had left the church that I was consciously thinking about the earth's history and humanity's origins, and I couldn't stop reading. I don't think I had ever been exposed to that much information about evolution at once in my whole life. I was at the park with a man whom I was crazy in love with, but I wasn't self-conscious about my ignorance. On the contrary: by that point in my life, I had learned that it was better to readily admit when I didn't know something, so I could learn it right away. That's what I did, and I hoped he would think better of me for it, but I'm not sure whether he did. I don't recall him making much of a reply when I told him I the information that I was reading there was the first time I'd never learned any of that stuff before.

(Elizabeth) In your personal opinion—do you believe that there are others in situations, who have had similar doubts but chose *not* to voice these?

(Kayla) Absolutely. In the church, you're really taught to not think for yourself, and if you're taught the same thing at home and at school, you don't ever learn to think critically or challenge inconsistent theories. Without that basic skill, you don't know how to challenge what people in your community believe wholeheartedly. On top of that, it's very counterintuitive to challenge anything, because what you've been taught all along is that these points are incontrovertible. I think people who are raised in isolation from other communities, as I was, are at a particular disadvantage. I would say the same even about kids who are raised in Christian schools, which is what I see happening with my nieces. If you're never exposed to other lifestyles and theories, how can you possibly develop enough interest in them to pursue even learning about them? To overcome all these setbacks, I think it takes a person who is rebellious and obscenely curious by nature. May all such people be so lucky.

(Elizabeth) Do you have anything else that you would like to add? Maybe about how you see evolution and religion now.

(Kayla) I honestly don't think much about the origins of life. It makes me uncomfortable to do so, which I assume is due to residual pressure from the theories so heavily impressed upon me for a solid twenty of my formative years. So, I just don't think about it. I believe in dinosaurs, and I mourn that I'll never get to see one. I can admit that probably no human ever saw one, but that's about where my science stops. I just don't think about it. I had to rewire so many parts of my brain once I left the church that it became exhausting. I suppose I got to the origins of life and just quit. I was tired of trying to 'fix' all of the things that I had been taught. I just don't think it matters. I was more concentrated on un-programming other things, like becoming less judgmental of other people and their lifestyle choices. Maybe it's simplistic, but I rationalize my refusal to think about where we came from by arguing that it doesn't matter nearly as much as where we're going. Most days, I even believe that, but the outlook is getting depressing enough that I'm finding myself more and more contemplating past epochs in human history. Maybe that will eventually lead me to the days before humans and apex predators who are too crafty for their own good.

When it comes to religion, I have a great many negative things to say against religion. I condemn religions for marginalizing women, for teaching that blind faith is preferable to deduction and reasoning, for turning the world into black and white and ignoring all the shades of gray, for allowing followers the surety of a get-out-of-jail free card—thereby freeing them from accountability to their fellow people—for insisting that non-believers will burn in hell for all eternity, and for a great many other wrongs.

I always want to be able to say that I now believe religion to be a great evil, but I can't, because I have very true friends who are really devote and really good

**Figure 27:**  Kayla at the Natural History Museum in London.

friends and good people. So, religion doesn't produce all bad eggs. So, what I try to do, especially with my relationship to my mom, is to respect people in a sense like sovereign nations. They have a right to believe whatever they want, if it is going to make them better and stronger. But I know for me, it did not make me a better person. I was mean, I was a cruel Christian, I judged people. I was rude to people because they were different to me and because they weren't Christians. And I don't want to be like that again. So, I don't want to say that religion is bad because to make that accusation would be another way of setting up my own world of black-and-white version of the world. I know that way, it is easy, but I am now very wary of any such seductively easy answers. I've known good and bad people in both camps, so all I can do is my best to respect people's sovereign right to adhere to ideas and principles that make them strong and good. Religion made me strong, in a way, but it definitely did not make me good. While I was religious, I practiced all the misinformed principles I listed above, and more, and so I believe that it is the wrong path for me. In the end, I think what really matter is whether or not we are being kind to one another. That is where I want to invest my energy—to become kinder, more creative and the best version of myself for me, for my partner and well, the rest of the world.

## In Search of Our Common Humanity

*"I am inspired by the notion of a responsibility to a universe that is governed by something bigger than me. It's okay for smart people to believe in religion… . There's a force in the universe that underlies all of this beautiful chaos. And understanding the relationship between science and God, I think, makes me a better scientist and a more complete person."*

Mayim Bialik

The division around evolution has made it too easy for each side to villainize the other. My conversations with Pete and Kayla changed the way I understood a creationist upbringing and helped me understand the complexity of this belief system from a psychological and social perspective. The reproduction of my conversations with them should act as a means of resolving the trance of the unreal other in order to 'rehumanize' those individuals who have become face-less entities in this century-old conflict. Hearing their personal accounts of growing up in creationist circles, the transmutation of their views on fundamentalism and their ultimate reasons for abandoning their creationist beliefs allow us to better understand the very complex emotional nature of creationism and to recognize the means by which individuals can be supported in their movement away from fundamentalist beliefs—namely through kind social contact and personal connection.

While it is easy to recognize how secularists, new atheists, creationists and fundamentalists differ according in their worldviews and belief systems, at our core, we all have many more traits, experiences and characteristics in common than those which divide us into our neat categories. This common humanity and intrinsic connection need to form the basis of our educational and communication approach. Effective teaching and science communication require much more than glossy illustrations and well-worded facts. We have to remember that these words and lessons are being transmitted to other humans and that as humans we are inherently social. As we have seen, a boomerang effect is most likely to occur when (Hovland et al. 1953):

(1) Perceived weak arguments come from negative sources [e.g., unclear information on evolution from teachers whom the students do not like or do not trust].

(2) The audience believes the speaker is trying to convince them of a different position than what the speaker claims [e.g., the teacher claims that they want to teach about evolution, but the students believe that the teacher convince them to become atheists].

(3) The communicated message triggers negative, aggressive or unremitting emotional arousal [e.g., lessons on evolution that cause fear, guilt or anxiety].

(4) The communicated message increases the recipient's knowledge of the norms and increases their conformity [e.g., making creationists believe that religion and science are inherently incompatible thus causing them to choose].

(5) Non-conformity to their own ingroup results in feelings of guilt or fear of ostracization [e.g., fear of losing connections to religious community by accepting evolution or questioning fundamentalist teachings].

(6) The speaker's view point is too different from the recipient's view point, thereby causing a 'contrast' effect and an enhancement of the recipient's original beliefs [e.g., the student sees the teacher as an unreal other or vice versa].

As we can see, divisive commentary made by new atheists only encourages the occurrence of a boomerang effect, meaning that instead of encouraging conceptual change—we are actually supporting belief perseverance. However, we could potentially pave the path towards conceptual change by:

(1) Pairing strong arguments with positive sources [e.g., offering better lessons from trusted teachers].[27]

---

[27] The terms 'teacher' and 'student' can also be understood as any form of social dynamic where a person is trying to communicate the validity of the theory of evolution to a person who does not yet fully understand it or actively rejects it.

(2) Dispelling fears that the speaker has any ulterior motive [e.g., that students trust that lessons on science are not actually attempts to dislodge their religious faith].

(3) Communicating messages in a way that triggers positive emotional states [e.g., information on evolution that trigger curiosity, wonder and interest].

(4) When the communicated message decreases the recipients' desire for conformity [e.g., removing the idea that one needs to decide between their religion and science].

(5) Helping the recipient deal with their feelings of guilt and fear of ostracization due to non-conformity to their own ingroup [e.g., validating and respecting the fear that the creationist has of losing important social connections by accepting evolution or questioning fundamentalist teachings].

(6) Avoiding the 'contrast' effect by focusing on commonality rather than dissonance [e.g., classroom or societal interactions that focus on common humanity instead of division over worldviews].

While new atheists allow themselves to vilify religion and creationists, all this accomplishes is an increase in the perceived battle-lines and thus the likelihood of conceptual conservatism. Our best chance of increasing scientific literacy lies in the hands of teachers and science communicators, but their work is often incumbered by the antagonistic commentary made by new atheists as these negative comments can be forwarded, copied, posted, retweeted and turned into memes across the globe. The replication of these hostile statements feeds into the creationists' fears and they become convinced that all science educators have ulterior atheistic motives. They become distrustful of science communication and science education believing that these can be equated with a movement to eradicate religion. So, in order to support an increase in global scientific literacy, we need to create an environment where it is easier for teachers to teach and for students to learn. Overall, we need to decrease sentiments of division and increase interest in science so that communication about scientific data is received by a receptive audience. One means of encouraging this type of receptivity is through clear, positive communication of data using the proposed communication guidelines (Table 14).

I have aimed almost my entire communication in this book at educators, science communicators and scientists, calling for them to develop their understanding of this issue. I have done this not because I think that group of individuals represent the problem, but because I am aware that teachers and scientists are much more likely

**Table 14:** Proposed guidelines for science educators and science communicators.

| Guidelines for effective science communication and science education |
|---|
| • I recognize that the goal in both science communication and science education is to promote interest and curiosity in science—not an aversion to it. |
| • I will internalize the concept of common descent and promote unity, understanding and tolerance in place of close-mindedness, intolerance and division. |
| • I will exemplify the curiosity, open-mindedness and tolerance that I wish to awake in my audience by abstaining from divisive speech, prejudice and intolerance. |
| • I acknowledge that my ultimate goal is to increase scientific literacy and that if I pit science against religion, I may actively encourage the faithful to reject science. |

to read this book than a fundamentalist or church leader (at least that is my opinion). Yet, despite this belief, I would like to add an appeal to fundamentalist leaders here to also assist in bringing this battle to a close. While pitting science against religion may cause some to uphold their original fundamentalist belief system, the rapid advancement of science and the abundant availability of scientific information makes it very difficult for such individuals to maintain their literalist beliefs over the course of their lifetime and reduces the chances of convincing future generations of such ideas as a young Earth. As we saw in the conversations with Pete and Kayla, as students leave their parents' homes and begin to have contact to individuals with new and opposing viewpoints, a natural process of reflection occurs. This reflection and subsequent doubt in fundamentalist teachings can result in the entire loss of a faith system. As we heard from Pete, "Given my background though, I couldn't bend so I broke." Kayla expressed similar feelings saying, "When I finally broke fully from the church, I had entered a new realm of 'sick-to-death of the hypocrisy', so it was easier to be apart than a part of it."

While I am not a religious leader or teacher, I can imagine that such a loss of faith is not the desired effect of fundamentalist teachings. For that reason, it is in the best interests of religious teachers and fundamentalist leaders to also abstain from pitting religion against science as proposed by the clergy letter (Figure 2). Instead of promoting division and separation, such leaders should be able to communicate their religious messages kindly and ethically. For this purpose, I would generally suggest that religious teachers (1) actively teach the differences between faith and science so that the faithful abstain from searching for signs of God in the mechanical details of the flagellum of a bacteria; (2) encourage members of religious communities to work towards developing tolerance, respect and support for the teachers who have been entrusted with the education of our children; and above all (3) abstain from promoting literalist interpretations of the Bible. While emphasizing a literal interpretation of the Bible causes adherent to reject modern science, at some point an understanding of science may cause an adherent to doubt *all* religious teachings that were based on the concept of literalism and subsequently suffer an entire loss of faith.

At the classroom level, we can also promote an understanding of common humanity and thus increase feelings of connection and trust among classmates through concrete exercises. For a long time, I was also convinced of the irreconcilable differences between secularists and creationists, scientists and the religious, fact seekers and blind believers. I was also under the 'trance of the unreal other'. My conversations with Pete and Kayla opened my eyes and I think that we can actively encourage this type of eye-, mind-, and heart-opening experience in classrooms through a guided exploration of common humanity. In theory, students can actively develop a sense of commonality by sitting together with other students and going through a series of specific questions that is designed to identify how different backgrounds lead to different viewpoints and to help participants identify what they have in common regardless of their differences in opinion. The ultimate goal of this exercise is to realize how we are all influenced through our upbringing, culture, religious beliefs and personal experiences and to recognize that these differences in worldviews are not what define us has human beings. Our varying worldviews are simply the result of varying personal histories and should therefore not define us or limit us. Regardless of whether someone is a democrat or a republican, a creationist or a new atheist, there are often many ideas, values, interests and concerns that these individuals have in common. It might be as simple as a love for dogs or concern for a sick relative, the point is to encourage a conversation that allows seemingly very different individuals to identify what unites them. A full description this exercise can be found in the Appendix V.

The path to compassion necessarily passes through a state of understanding and empathy. The miscommunication between secularists and the religious is based on a severe lack of understanding and empathy. Each side is fearful, and yet neither side is capable of understanding the other side's fear. On the one side you have scientists and secularist thinking 'How can anyone be afraid of science?' while the evangelical is equally confused wondering 'How can anyone be afraid of students praying in school?' Yet, the fear for each side is real for that individual.

Observing the lack of understanding between secularists and creationists is akin to watching the communication between a dog and cat, where neither can understand the fears or perspectives of the other. While a dog may reprimand the cat for being afraid of cucumbers, the cat might be flabbergasted by the dog's incessant need to seek the approval from humans. Clearly the dog is incapable of understanding the cat's fears of cucumbers are a product of its environment and the conditions in which their ancestors lived (possibly related to perceived danger of snakes) and the cat is equally unable to understand the dog's need for companionship. In the same way, we as humans tend to judge the emotional reactivity of another individual as irrational (insane, ignorant, wicked, etc.) when we are incapable of recognizing that this reactivity is also a product of upbringing and social context.

The good news is that humans differ from other animals in this context in two major ways: (1) While the dog will never be able to understand or empathize with the cat's fears, we as humans are able to develop this level of empathy through the engagement of our higher cognitive functions and increased understanding for the context and roots of those fears, and (2) while the cat will not be able to un-learn its fear of cucumbers in its lifetime, I am certain that with purposeful, positive teaching humans can overcome their fears, especially their fears of the 'unreal other'.

## Summary

According to the theory of common descent, we are all intrinsically connected and yet even the most ardent promoters of evolutionary theory (e.g., Coyne and Dawkins) overlook this basic tenet. Instead of recognizing our common humanity, new atheists and fundamentalists continue to promote a primitive dichotomous view of world where people of science and people of faith are forever at odds with one another. These defensive and offensive communications from both sides are the driving force behind this century-old conflict. Ironically, the louder and more aggressive each side argues, the more ineffective their communication becomes. In other words, the more new atheists *demand* that the creationists give up their long-held views, the less likely this is to ever occur. The only true means of encouraging conceptual change and increasing scientific literacy, is by relying on calm, logical, kind discussions that give an individual the time and space required to reflect, digest and then to either accommodate or assimilate the new information to their preexisting conceptual structures. In order to participate in this type of conversation, it is imperative for all parties involved to develop a better understanding not only of the complexity of this topic but also the position of the other conversation partner. Or as Pete said, "Like a chess player walking around the board to see his opponents view of the game, effective persuasion is rooted in a deep understanding of the audience."

# Conclusion

*"Those who tell the stories rule society."*                                              Plato

For many it may still be incomprehensible that around 40% of adult Americans continue to believe that the Earth is less than 10,000 years old. While many assume that this belief is a sign of ignorance, what we have seen through the analysis of the creationist phenomenon from historical, cultural, legal, psychological, evolutionary, neurological, educational and societal perspectives, is that these beliefs are actually the product of a conscious decision to uphold fundamentalist views while actively rejecting all scientific data that contradict these views. The fact that a person would intentionally perpetuate such a false understanding of the natural world may be even more unfathomable for some than the beliefs themselves. While the ignorance-assumption presumes that if these individuals were exposed to the correct data that they would be capable of amending their misconceptions, this is simply not the case. The fact that these beliefs are intentionally upheld highlights not only the complexity of this issue but also underscores the tenacity of these beliefs and explains why they are often immune to educational interventions.

At the center of this belief system is the literal interpretation of the Genesis story of creation. As young-earth creationist leader Ken Ham claims, the creation story forms the foundation of Christianity and if the story of Genesis were to be lost then all of Christianity would tumble (2013). This idea is not only put forth by creationist leaders but also lived by creationists as we learned from Pete who told us that the literal interpretation of the Bible forms the basis of the creationist house of cards, "If you give yourself the authority to remove something uncomfortable here or there, then everything is up for grabs... . Your community, your family, your marriage, your traditions, your morals all collapse if you remove one card, such as the Genesis story." This emphasis on the literal interpretation of the Bible is the heart of the creationists' fear of evolution as evolution does directly contradict the *literal* interpretation of Genesis. This focus on the literal interpretation of the Bible is also the main argument for why creationist beliefs should be categorized as fundamentalist beliefs rather than religious beliefs.

While a literal interpretation of Genesis is unequivocally false in terms of understanding the natural world, these fundamentalist beliefs offer creationists certain psychological advantages as they support their ability to manage difficult emotional states while simultaneously cultivating positive emotional states because these beliefs act as the basis for their meaning system, terror management strategies and social network. The clear black-or-white world of good versus bad which is created through fundamentalist beliefs, for example, allows the creationist 'to make sense' of the chaotic world around them—in essence creating order from chaos, while the belief in an afterlife provides a means of reducing death anxiety. The clinging to these beliefs can thus be understood in an evolutionary context as it appears to be driven by the our innate (evolutionary-driven) desire to avoid pain and seek pleasure. While this general behavior assisted our primitive ancestors' survival by encouraging them to seek shelter, mates and food while avoiding dangerous life-threatening situations, these behavioral drives can be maladaptive in modern society.

Another possible evolutionary-based explanation for creationist clinging is our inherent human need for social connection. Historically our survival was closely linked to our tribal existence where our access to vital resources such as food, mates and support was largely dependent upon the degree to which other tribal members valued one's welfare (Sznycer et al. 2016). Humans have thus adapted over millions of years not only to recognize the value of human relations but to also guard against any behavior that could cause expulsion from familial groups or other community structures. This need for human connection and the recognition of the value of particular social connections thus offers a further explanation for why creationists cling to their beliefs as these beliefs act as a key to a particular community of like-minded individuals. Recent research has even brought forth evidence showing that these religious (or fundamentalist) communities offer measurable psycho-social and physical benefits to members.

This link between community and creationist beliefs became very apparent through my conversations with Pete, who described the difficulty he had in coming to terms with his doubts about the literal interpretation of Genesis because he knew that admitting or voicing his doubts could cause him the loss of numerous friendships. At the same time, we saw that the relationships within these fundamentalist communities can be very brittle as they are based on a literal interpretation of the Bible and are thus associated with a strict adherence to a specific set of fundamentalist rules. As we heard from Kayla, these relationships can be easily lost once a person engages in activities that are considered 'un-Christian' such as participating in a school theater production or cutting your hair.

By recognizing how creationist beliefs form the basis of so many psychological coping strategies and social networks, it becomes more comprehensible why creationists cling to these beliefs and also allows us to understand why they are so keen to reject evolution. While there is not an intrinsic conflict between religion and science, there is a clear conflict between the literal interpretation of Genesis and scientific data as it is impossible for the Earth to simultaneously be 10,000 years old and 4.5 billion years old. What this means is that when a person's faith is based on the literal interpretation of Genesis, exposure to scientific data regarding the natural origins of man could in fact cause confusion or doubt to arise.

Once these belief systems begin to totter, the creationist's meaning system, terror management strategy and access to valued communities become endangered and ultimately the entire fundamentalist house of cards can come tumbling down. It is then clear why evolution is perceived as a threat to this group of believers as it does in fact have the potential to cause a major disruption in a person's inner- and interpersonal life. Recognizing how and why evolution is perceived as a threat allows us to understand a creationist's resistance to evolution as a byproduct of our own evolutionary past because it is driven by behavioral tendencies that helped prehistoric humans survive, e.g., avoid pain, seek pleasure and guard against expulsion from our tribe.

We have also evolved to be able to quickly recognize and respond to threats in our environment. It is thus not surprising that when a person feels like their faith, and ultimately their chance of eternal life and access to community are at risk, the perception of evolution as a threat can become so strong that it can cause a fight-or-flight response. In this way, we can see creationists' negative reaction to evolution as a mutated form of 'struggle for survival' as they appear to be fighting for their eternal lives as they perceive the teaching of evolution as a threat to their chance of salvation (or that of their children). Likewise, they also appear to be defending their 'tribe' against attack from other 'tribes' (secularists) in an attempt not only to protect the tribe itself but also their personal membership to that tribe.

Here it must be made very clear that understanding the psychological and evolutionary roots of these behaviors does not mean that the goal is to encourage individuals to maintain these beliefs. It is important, however, to comprehend how and why negative emotional states arise in creationists in response to evolution so that we can prevent or at least mitigate those emotional states instead of triggering or exacerbating the fight-or-flight response. This is important not only socially but also educationally as this fear-based response renders the individual incapable of absorbing or processing any data presented to them.

The role of emotions in the discussion of human origins cannot be ignored or underestimated because there is a clear relationship between emotions and cognitive ability, whereby negative emotional states are associated with a narrowing of the cognitive functioning (Powietrynska et al. 2014). Within a classroom setting, these negative emotional states occur most frequently when a person's convictions are challenged by scientific findings and negative reactions to these data are highest when a person's identity is involved in the original conviction, which is clearly the case for many creationists.

If we ignore the emotional nature of this topic, we become not only ineffective as educators and communicators but also increase the likelihood of a boomerang effect or conceptual conservatism. Ultimately, effective science communication requires a receptive audience and the prerequisite of this type of receptivity and cognitive flexibility is a shift into more neutral emotional states. By understanding the evolutionary causes for creationist behavior, it should be easier to understand the source of their anxiety and why certain individuals are so resistant to learning about evolution.

Recognizing this degree of resistance and anxiety is particularly important when we begin to look more extensively at learning and the remediation of misconceptions as the process of conceptual change is reliant upon an individual's own willingness to engage with the data presented to them. This willingness is often lacking in creationists because the degree of conceptual restructuring that is required to go from the idea of a young Earth to an ancient Earth necessitates not only a new understanding of the natural world but also brings with it a reformation of one's self- and world-theory. This is due to the fact that for many creationists, the story of creation forms the foundation of their identity and thus a change in the understanding of human origins ultimately leads to a change in the way they see themselves and their role in the world. A shift in the understanding of human origins can thus cause them to suddenly question—*Who I am? What is my role and purpose in the world?*

So, when creationists are presented with evidence for evolution, it is more than just an educational intervention, it is a direct challenge to deeply seeded self- and world-theories, and it is therefore not surprising that they react negatively or are non-responsive as they attempt to guard against the ultimate destabilization of their meaning system. In other words, while teachers might be simply explaining how mutations can give rise to new species over time, the creationist students are experiencing an entirely different situation. For them it is not a simple classroom discussion, but instead a presentation of information that may cause them to doubt the story of creation which has served as the bedrock of their faith.

It is thus necessary to rethink the manner in which we communicate and teach about evolution in order to assist creationist students in a movement away from conceptual conservatism and towards conceptual change. The only way that we prevent a fight-or-flight reaction is by removing the perceived sense of threat around science and evolution. This will require new educational approaches that allow us to have an authentic dialogue with creationists in order to encourage interest and receptivity to science.

The two major steps that are required to obtain this goal are (1) a reduction in the degree of division that occurs at an individual and societal level around this topic and (2) an increase in cognitive flexibility on all sides. Step one requires us to first recognize the difference between faith/spirituality/religion and fundamentalism. Although, new atheists and fundamentalist leaders adamantly argue that *religion* and science are incompatible, the only real incompatibility is between *fundamentalist* thinking and science due to the emphasis that fundamentalists place on the *literal* interpretation of sacred texts. While different creationist groups place a varying degree of importance on the literal interpretation of the Bible (Figure 8 and Figure 9), the concept of special creation is central to all creationist variations and is thus a direct contradiction to organic evolution. Making this differentiation between religion and fundamentalism is crucial as we want to discourage dogma, ideology and fundamentalist thinking, yet we do not want to try to dissuade anyone from their own personal belief systems, faith or spirituality. In other words, we are not trying to replace one dogmatic belief with the other but simply encouraging an overall increase in open-mindedness.

We must keep in mind that the goal of science communication is not to scare people away from science. Instead, we want to encourage the development of positive feelings about evolution and science, not only to increase the chance of progressive scientific literacy but also to encourage more young people to pursue careers in the sciences. This requires us to abstain from divisive speech, diatribes and condescension. Aggressive, insulting speech only causes the religious to be weary of science. Likewise, ridicule and criticism have never been the foundation of fruitful dialogues and ultimately, if we want to increase general scientific literacy, we need an audience who is willing to listen, engage with the data and none of this is possible when the intended audience is feeling attacked and defensive or is engaged in a fight-or-flight response.

The basis of a fruitful conversation is instead rooted in both parties' ability to engage with their so-called upstairs brain. Much of the conflict surrounding evolution can be traced back to primitive structures in our downstairs brain that originally developed to help our ancestors survive in hostile prehistoric environments. While these behaviors (e.g., pleasure seeking and threat recognition) were crucial for survival in the past, they can be profoundly maladaptive in our modern society. This is true not only when it comes to nutrition and daily stresses but is particularly true with regard to the conflict between creationists and secularists. One means of breaking out of this vicious cycle is by making a conscious effort to act and communicate using our higher cognitive faculties. While our mind contains the remnants of our prehistoric past, it also possesses a seemingly endless potential for change in the form of neuroplasticity.

While we cannot undo years of cultural programming, we do have the ability to change and shape the way our brain reacts to certain triggers. In fact, our primitive threat recognition system can actually be overridden by activating and intentionally training more highly evolved portions of our brain. This type of self-directed neuroplasticity is possible do to the malleable nature of the neural circuitry in the neocortex. One means of engaging in self-directed neuroplasticity is through the practice of mindfulness. Mindfulness as a form of contemplative practice supports the development of self-regulatory skills associated with the ability to become aware of one's internal environment (e.g., emotions, judgements, beliefs, thoughts) and consciously choose how to react to these mental productions.

Learning to become aware of one's own thoughts, judgments, emotions, actions is a major step towards conscious gear-shifting as much of social life normally occurs without conscious awareness. When we are not aware of how we are feeling, we tend to react to situations based on long established behavioral patterns. In other words, the more often we engage in certain behaviors, the faster and more likely we are to act in the same manner each and every time we are exposed to similar stimuli—ultimately developing into a fully-automated response. While automated responses help us in our daily lives to perform certain physical processes such as driving a car, this type of automation can negatively affect our interpersonal responses as it often limits the degree of flexibility we display in response to other individuals in our environment. This is particularly true in the case of discussions regarding human origins where automated responses often prevent effective dialogues from occurring between creationists and secularists.

For example, let us imagine a person (Individual A) who is convinced that evolution is a dangerous (possibly evil) idea and that 'believing' evolution will cause them to lose their faith and their membership to a particular religious community. Each time Individual A encounters another individual who attempts to explain the data that support the theory of evolution, Individual A becomes fearful and defensive, immediately shutting down and arming themselves with counterarguments. The more often Individual A engages in this defensive behavior, the quicker they are to march down the proverbial war path as soon as someone even mentions the word evolution.

Likewise, we can image another person (Individual B) who acts in disgust as soon as someone even says they are unsure about whether evolution is true and automatically judges such individuals as being ignorant. Based on this knee-jerk reaction, Individual B would automatically speak to Individual A with distain and condescension. Clearly, the conversation between Individual A and Individual B would be ineffective and both would leave the table with their original belief system, opinions and fears not only still well intact, but also validated.

Here it is clear that a change in mindset and behavior is necessary for these two individuals to have a real conversation. If Individual A and B can, for example, become aware of their emotional state and the thoughts driving their practiced response, they could slowly develop the ability to choose how they want to act in a particular situation. They could then decide, for example, whether or not they want to let inner Neanderthal drive their reactions and engage in a heated confrontation or if they want to engage in a conversation using their more evolved brain. Awareness allows us to *respond* instead of just *reacting*.

Mindfulness practices are thus particularly relevant to the topic of creationism and evolution as it supports more effective communication practices and educational interventions as practitioners of mindfulness are capable of increasing their cognitive flexibility and social awareness. For example, when mindfulness is developed as a class, students and teachers are better able to recognize not only how they are affected by their environment but also how they affect others in their surroundings. Once a certain degree of mindful awareness and reflectivity has been achieved, individuals are better able to rationally change aspects of their behavior to the benefit of themselves and others. This movement towards more conscious action allows for the development of a more compassionate and less judgmental stance in both general science communication and formal education. In this way, we are able to create an environment where creationists are less likely to feel attacked and thus, we can reduce the likelihood of an automatic fight-or-flight response. Once creationist students become more responsive to science, mindfulness can act as an important tool in addressing particular misconceptions. In this particular situation, mindfulness is analogous to intentional conceptual change in many ways as it reflects a voluntary state of mind that connects motivation, cognition and learning and can therefore support a person's ability to accommodate or assimilate newly acquired information (Duit and Treagust 2003; Salomon and Globerson 1987). Lastly, mindfulness can help students manage difficult emotions that may arise as they restructure previously held conceptual structures and are subsequently faced with a destabilization of their meaning system or terror management strategy.

In the end, we need to keep in mind that our ultimate goal is to increase scientific literacy but that this process should not cause any undue emotional suffering or harm. For that reason, we will need to refrain from perpetuating a divisive and dichotomous view of the world. Our promotion of science should thus be framed by open-mindedness and tolerance. While a world without science is a poor world indeed, for many individuals, a life without faith is a life void of meaning.

It is therefore time for us all to embrace the tenet of common descent and recognize our inherent interconnectedness. By understanding our own inner Neanderthal, we can learn to bid goodbye to our primitive tendencies as we rely more heavily on our more highly evolved brain regions. In recognizing our ability to consciously alter our neural pathways and thinking patterns through mindful decisions, we are able to use our understanding of evolution and biology to develop better educational and communication practices that encourage the growth of scientific literacy. At the same time, this movement away from our inner Neanderthals allows us to become the active drivers of our own conscious evolution and to lay the groundwork for a society based on connection rather than division.

# Afterword

While this book has focused on fundamentalist opposition to evolution, we must be aware that evolution is not the only scientific fact that is actively opposed within our current society. In the United States, there is also a fervent denial to climate science. This is seen most clearly by the fact that Americans rate amongst the most educated about climate change but fall below the world's average with regard to accepting that climate change is caused in large part by human activity (Pelham 2009).

This type of science denial is not only reminiscent of evolution-denial but according to certain studies there appears to be a solid link between the two ideologies. A PEW study from almost ten years ago examined the link between religious thought and perception of scientific data and the results of this study showed that there was a linear relationship between anti-evolutionary thought and climate change denial (Lugo et al. 2008). This PEW study specifically found that almost all evangelical groups are opposed to evolution and environmental regulations, while Buddhists and non-Orthodox Jews, for example, are both in support of pro-environmental regulations and have a high acceptance of evolution. One of the only true outliers were the Jehovah Witnesses, who are in support of environmental regulations yet exhibit a high rate of denial of evolution (Figure 28). The PEW findings were graphed by Josh Rosenau, allowing the linear trend to be more easily recognized (Rosenau 2015).

The implications of this connection are a major cause for concern. While a well-orchestrated creationist movement has devoted the last 100 years to actively trying to keep evolution out of the classroom, a new trend appears to be moving towards the joint opposition to teachings on evolution and anthropogenic climate change, meaning that the same legally savviness that the creationists have used to try to keep evolution from being taught in the past is now being applied in an attempt to students from learning about climate science. While the prevention of lessons on evolution hinders a student's general scientific literacy, the hinderance of lessons on climate science could be even more devastating as scientists all over the world continue to produce data which highlight the necessity of joint global action in the mediation of the problems our modern lifestyles have caused on the natural environment.

The joint effort to undermine both evolution education and climate science education is seen most clearly in a suit that was brought to court by the Citizens for Objective Public Education (COPE) against the Kansas State Board of Education (SBOE). COPE filed this suit in hope of reversing the Kansas SBOE's 2013 decision

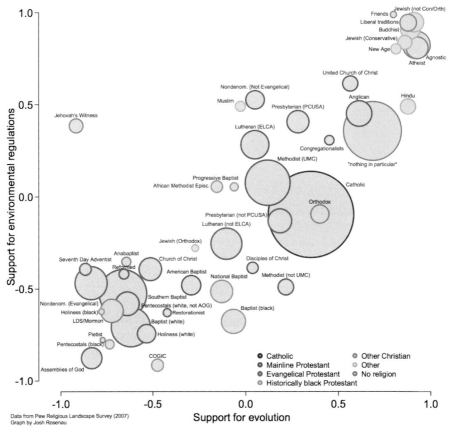

**Figure 28:** Graph by Rosenau illustrating how anthropogenic climate change denial and evolution opposition coincide, while as acceptance of evolutionary theory increases, so does support for environmental regulations.

**Color version at the end of the book**

to adopt the Next Generation Science Standards (NGSS). They claim that the standards endorse a non-theistic worldview. Although the majority of the complaint revolves around the typical creationist arguments concerning the teaching of organic human origins, they also contend that teaching impressionable students about environmentalism is in violation of their constitutional rights.

This can be seen most clearly in Exhibit B of the original complaint, which contains a letter from the president of COPE, Robert P. Lattimer. As Lattimer explains in that letter, "The Framework and Standards seek to imbue the students with a particular view regarding the manner in which humans should respond to climate change, sustainability, and other environmental matters". As part of Exhibit B of the original complaint and Appendix D of the appeal to the Supreme Court, Lattimer also included a list of the specific NGSS statements which COPE objects to (Table 15).

**Table 15:** Ideas about the environmentalism and sustainability included in the Next Generation Science Standards, which COPE president Lattimer claims are in violation of students' constitutional rights.

| Section | Specific Statement |
|---|---|
| HS-LS2 (grades 9-12) | L S2 C: Ecosystem Dynamics, Functioning, and Resilience<br>…Moreover, anthropogenic changes (induced by human activity) in the environment—including habitat destruction, pollution, introduction of invasive species, overexploitation, and climate change—can disrupt an ecosystem and threaten the survival of some species. (HS-LS2-7) |
| | LS4 D: Biodiversity and Humans<br>…But human activity is also having adverse impacts on biodiversity through overpopulation, overexploitation, habitat destruction, pollution, introduction of invasive species, and climate change. |
| HS-ESS2 (grades 9-12) | ESS2. A: Earth Materials and Systems<br>…The geological record shows that changes to global and regional climate can be caused by interactions among changes in the sun's energy output or Earth's orbit, tectonic events, ocean circulation, volcanic activity, glaciers, vegetation, and human activities. (HS-ESS2-4) |
| MS-ESS3 (grades 6-8) | ESS3. D: Global Climate Change<br>Human activities, such as the release of greenhouse gases from burning fossil fuels, are major factors in the current rise in Earth's mean surface temperature (global warming) …. |
| HS-ESS3 | ESS3. C: Human Impacts on Earth Systems<br>The sustainability of human societies and the biodiversity that supports them requires responsible management of natural resources. (HS-ESS3-3) … |

As Lattimer explains, these topics impact "not only religion, but also political and Constitutional views regarding human liberty, the right to property, and the proper role of government" (COPE's original complaint against the Framework & Standards-Exhibit B/Appendix D COPE's appeal to Supreme Court 2016). As Lattimer describes the problem, the HS-LS2 part of the Standards focuses on "ameliorating the negative effects of human activities—without giving consideration to the negative effects of governmental regulation on human liberty (2013)". Lattimer also contends that the HS-ESS2 and MS-ESS3 portions of the Standards require more explanation that helps students understand that "there is widespread debate among climate scientists over (a) the extent to which greenhouse gases (GHG) contribute to changes in global temperature, (b) the degree of climate sensitivity to atmospheric carbon dioxide, (c) whether the consequences of GHG warming will be net beneficial or net harmful, and (d) whether the benefits of any attempts to reduce GHG emissions would be worth the costs (Lattimer 2013)."

Regarding section HS-ESS3, Lattimer has a particular problem with the idea of sustainability. While he admits that the "general idea of protecting the environment and conserving natural resources is not controversial," he does argues that 'sustainability' is an ideology and not science as it has become a political movement emphasizing "simpler lifestyles, reduced economic development, global redistribution of wealth, limited use of natural resources in developed countries, 'green' (renewable) energy, 'smart growth' policies, human population control, and global governance (Lattimer 2013)." As he concludes, he questions the wisdom of raising such ideas with impressionable students and argues that the NGSS coverage of environmental issues lacks objectivity.

While the wording and general legal strategy seen in this exhibit already provide a clear connection to earlier creationist legal maneuvering, a closer look at the biography of Robert Lattimer illustrates his own personal ties to the creationist movement. On his resume, he appears to be a successful and knowledgeable scientist, with a B.S. and PhD in chemistry and the proud founder of Science Excellence for All Ohioans (SEAO). Yet, upon closer inspection, the organization that Lattimer founded, Science Excellence for All Ohioans, turned out to be a pro-intelligent design establishment that was involved in petitioning for changes to the science standards in Ohio, claiming that the biggest problem in science education in Ohio is the 'censoring' of 'evidence for design'. In 2000 they made it their mission to lobby for changes in the newly introduced statewide science standards so that they would include intelligent design principles. The specific changes that the SEAO proposed to the new standards included the removal of a reference to the age of the Earth, the introduction of intelligent design as a scientifically valid alternative to evolution, and references to William Paley (*Natural Theology* 1802) and Michael Behe (*Darwin's Black Box* 1996) to be added where discussion of Darwin, Mendel and Lamarck take place.

Lattimer's link to intelligent design and other religious conservatives becomes even clearer through the awards he has received for his work such as the Eagle Award from the Eagle Forum and the Wedge of Truth Award from Intelligent Design Network. The Eagle Forum is an organization dedicated to helping conservative and pro-family men and women participate in self-government—they actively oppose gun control and support parental rights to the religious education of their children (http://www.eagleforum.org/misc/descript.html Accessed 24 September 2018). The Wedge of Truth Award is bestowed by the Intelligent Design Network to individuals for advancing intelligent design. Other winners of the Wedge of Truth Award include major leaders of the intelligent design movement, William Dembski and Michael Behe.

This complaint filed by COPE and the personal history of Lattimer clearly illustrates the link between creationism and climate change denial. Further examples of this link between the two movements can be seen through further actors who are invested in countering evolution and climate science such as the former Republican senator Rick Santorum. While Santorum focused his efforts on opposing evolution in 2001 with the proposal of the pro-intelligent design legislation the *Santorum Amendment*, ten years later in 2011 he went public with his views on climate change which according to him is an elaborate scheme that was concocted by the political left as a means of promoting greater government control (Santorum 2011).

Yet, it is not only climate science *deniers* who have taken a legal page from the creationist play book. After one hundred years of litigation aimed at creating a school system and political milieu that reflect fundamentalist values, the upcoming generation has their own ideas of how to use the legal system to promote their environmentally-conscious values. In response to nationwide climate-change denialism and US President Donald Trump's refusal to address the need for immediate action, a group of students has now filed a lawsuit against the US government, *Juliana v. United*

*States*, claiming that the US government has violated their "constitutional rights to life, liberty and property by failing to prevent dangerous climate change". Just as our organic origins cannot be denied, neither can our effect on the organic world around us. Whether or not this new trend is just the beginning of another century of legal strife will be decided by our ability as a society to communicate, resolve our differences and develop solutions that are truly in the best interest of all humans, regardless of our opposing political stances, worldviews or religious beliefs.

# Appendices

# Appendix I—Legal Conflicts

The following table offers an abbreviated overview of the plaintiffs, defendants and the general judgements made in each of the cases from 1925 to 2015.

**Table 16:** Overview of legal cases involving the teaching of evolution from 1925 to 2015.

| Court rulings between 1925 and 2015 | | |
|---|---|---|
| **Plaintiff** | **Defendant** | **Judgement** |
| Tennessee - state | John Scopes - teacher | Defendant found guilty of violating state's Butler Act. |
| Epperson - teacher | Arkansas - state | Any statute that prohibits the teaching of evolution is in violation of the Constitution. |
| Daniel - teacher | Waters - textbook commission | Genesis Act (requiring balanced treatment) is in violation of the Constitution. |
| Hendren - parent | Campbell - textbook commission | Including a textbook that discusses Creationism in public schools is in violation of the Constitution. |
| Segraves - parent | California - state | Teaching evolution cannot be considered the Establishment of Religion and is therefore not in violation of the Constitution. |
| McLean - reverend | Arkansas - board of education | Balanced Treatment Act (requiring balanced teaching of evolution and Creation Science) is in violation of the Constitution. |
| Aguillard - parent | Edwards - governor | Balanced Treatment Act (requiring balanced teaching of evolution and Creation Science) is in violation of the Constitution. |
| Webster - teacher | New Lenox - school district | A teacher's freedom of speech does not overrule a school district's obligation to protect a student's constitutional rights. |
| Peloza - teacher | Capistrano - school district | Requiring a teacher to teach evolution instead of Creationism is not a violation of that teacher's constitutional rights. |

*Table 16 contd. ...*

*...Table 16 contd.*

| Court rulings between 1925 and 2015 | | |
|---|---|---|
| **Plaintiff** | **Defendant** | **Judgement** |
| Freiler - parent | Tangipahoa - board of education | Requiring a disclaimer to be read before teaching evolution is in violation of the Constitution. |
| LeVake - teacher | ISD #656 - school district | A teacher's freedom of speech does not overrule a school district's obligation to protect a student's constitutional rights. |
| Kitzmiller - parent | Dover - school district | Requiring a disclaimer to be read and any attempt to include Intelligent Design into the classroom is in violation of the Constitution. |
| Hurst - parent | Newman - school district | The school district may not offer any courses promoting Creationism or Intelligent Design. |
| Selman - parent | Cobb - school district | A use of a disclaimer to warn students about evolution is in violation of the Constitution. |
| Caldwell - parent | Roseville - school district | It is not a violation of the Constitution if a policy proposed by a parent is not granted by a school district. |
| ACSI - school | Stearns - university | Rejecting courses from Christian private schools on the basis of scientific merit is not a violation of the Free Exercise Clause of the Constitution. |
| C.F. - student & parents | Capistrano - school district & teacher | It is not constitutionally permissible for teachers to make hostile remarks about religion or creationism in the classroom. |
| Comer - curriculum director | Scott - textbook commission | Neutrality clauses regarding employees' treatment of creationism and evolution are not in violation of the Constitution since they do not advance nor inhibit religion. |
| Caldwell - citizen | Caldwell - professor | Dismissed because defendant could not prove taxpayer status or concrete injury. |
| Doe - family | Mount Vernon - Board of Education | Settled outside of court. Financial reimbursement for the family and the teacher in question was fired. |
| ICR - creationist graduate school | Paredes - Texas Higher Education Coordinating Board (THECB) | Summary Judgement in favor of THECB due to lack of evidence brought forth by the ICR. |
| AFA - think tank | CSC - state science center | Settle outside of court. Neither party accepted any liability. CSC paid $110,000 in damages to AFA. |
| Hensley - teacher | Johnston County Board of Education | Plaintiff claims of constitutional violation were dismissed since free speech is not granted to teachers in an official capacity. |
| Lane - parent | Sabine Parish School Board Consent | Decree in favor of the plaintiff. The Board's actions of promoting religion if proved would be in violation of the Establishment Clause of the Constitution. |

The following are samples of the original text (statutes, policies, product descriptions, etc.) that played a central role in the legal rulings.

*From Epperson v. Arkansas 1968*

## Anti-Evolution Statute* (1929)

80-1627—Doctrine of ascent or descent of man from lower order of animals prohibited: It shall be unlawful for any teacher or other instructor in any University, College, Normal, Public School, or other institution of the State, which is supported in whole or in part from public funds derived by State and local taxation to teach the theory or doctrine that mankind ascended or descended from a lower order of animals and also it shall be unlawful for any teacher, textbook commission, or other authority exercising the power to select textbooks for above mentioned educational institutions to adopt or use in any such institution a textbook that teaches the doctrine or theory that mankind descended or ascended from a lower order of animals.

80-1628—Teaching doctrine or adopting textbook mentioning doctrine—Penalties— Positions to be vacated: Any teacher or other instructor or textbook commissioner who is found guilty of violation of this act by teaching the theory or doctrine mentioned in section 1 hereof, or by using, or adopting any such textbooks in any such educational institution shall be guilty of a misdemeanor and upon conviction shall be fined not exceeding five hundred dollars; and upon conviction shall vacate the position thus held in any educational institutions of the character above mentioned or any commission of which he may be a member."

*Initiated Act No. 1, Ark. Acts 1929. Ark. Stat. Ann. 80-1627, 80-1628

*From Hendren et al. vs. Campbell et al.*

## Product Description—Biology: A Search for Order in Complexity (Published by Creation Research Society in 1970)

Give your students a solid understanding of God's creation with this updated and improved Biology Text. Full-color photos, illustrations and charts throughout clearly display concepts discussed and help with visualization of terms and processes. From the scientific method to biochemistry to body systems and biogeography, each chapter looks at how God's plan and purposes are evidenced in creation. Scientifically accurate and true to a 6-day/young-earth creationism, Biology provides a scientific education that acknowledges God's role throughout. Review questions and suggestions for advanced study are included. Grades 10–12. 418 pages, hardcover, 2nd edition. www.christianbooks.com

*From Segraves v. California*

## California State Board of Education's Anti-Dogmatism Policy (1972)

The domain of the natural sciences is the natural world. Science is limited by its tools—observable facts and testable hypotheses.

Discussions of any scientific fact, hypothesis, or theory related to the origins of the universe, the earth, and life (the how) are appropriate to the science curriculum. Discussions of divine creation, ultimate purposes, or ultimate causes (the why) are appropriate to the history-social science and English-language arts curricula.

Nothing in science or in any other field of knowledge shall be taught dogmatically. Dogma is a system of beliefs that is not subject to scientific test and refutation. Compelling belief is inconsistent with the goal of education; the goal is to encourage understanding.

To be fully informed citizens, students do not have to accept everything that is taught in the natural science curriculum, but they do have to understand the major strands of scientific thought, including its methods, facts, hypotheses, theories, and laws.

*From McLean et al. v. Arkansas Board of Education et al.*

## Balanced Treatment for Creation-Science and Evolution-Science Act* (1981)

"Public schools within this State shall give balanced treatment to creation-science and to evolution-science." Section 4 Definitions, as used in this Act:

(a) "Creation-Science" is defined as scientific evidences for creation and related inferences that indicate: (1) Sudden creation of the universe, energy, and life from nothing; (2) The insufficiency of mutation and natural selection in bringing about development of all living kinds from a single organism; (3) Changes only within fixed limits of originally created kinds of plants and animals; (4) Separate ancestry for man and apes; (5) Explanation of the earth's geology by catastrophism including the occurrence of a worldwide flood; and (6) A relatively recent inception of the earth and living kinds.

(b) "Evolution-Science" is defined as being scientific evidences and related inferences that indicate: (1) Emergence by naturalistic processes of the universe from disordered matter and emergence of life from nonlife; (2) The sufficiency of mutation and natural selection in bringing about the development of present living kinds from simple earlier kinds; (3) Emergence by mutation and natural selection of present living kinds from simple earlier kinds; (4) Emergence of man from a common ancestor with apes; (5) Explanation of the earth's geology and the evolutionary sequence by uniformitarianism; and (6) An inception several billion years ago of the earth and somewhat later of life.

*Arkansas Act 590

*From Herb Freiler et al. v. Tangipahoa Parish Board of Education et al.*

## Original Disclaimer from Tangipahoa (1994)

It is hereby recognized by the Tangipahoa Board of Education, that the lesson to be presented, regarding the origin of life and matter, is known as the Scientific Theory of Evolution and should be presented to inform students of the scientific concept and not intended to influence or dissuade the Biblical version of Creation or any other concept.

It is further recognized by the Board of Education that it is the basic right and privilege of each student to form his/her own opinion and maintain beliefs taught by parents on this very important matter of the origin of life and matter. Students are urged to exercise critical thinking and gather all information possible and closely examine each alternative toward forming an opinion.

# Appendix II—CUNY Heuristics

The following tables include the full versions or excerpts from the heuristics developed by the CUNY group.

**Table 17:** Full version of CUNY's first mindfulness heuristic.

| | Characteristics in the First Iteration of the Mindfulness Heuristic |
|---|---|
| 1 | When I'm walking, I deliberately notice the sensations of my body moving. |
| 2 | I'm good at finding words to describe my feelings. |
| 3 | I do not allow myself to get distracted from the task at hand. |
| 4 | I don't criticize myself for having irrational or inappropriate emotions. |
| 5 | I perceive my feelings and emotions without having to react to them. |
| 6 | I have a hard time separating myself from my thoughts and feelings. |
| 7 | I am not curious to see what my mind is up to from moment to moment. |
| 8 | It is hard for me to put my beliefs, opinions, and expectations into words. |
| 9 | I do not feel the need to judge how I feel. |
| 10 | I seek to control unpleasant thoughts and feelings. |
| 11 | When I have distressing thoughts or images, they tend to consume me. |
| 12 | I rarely notice the wind in my hair or sun on my face. |
| 13 | I focus consciously on everything I do. |
| 14 | I am not curious about my thoughts and feelings as they occur. |
| 15 | When I'm terribly upset, no words can describe how I feel. |
| 16 | I make judgments about whether my thoughts are good or bad. |
| 17 | In difficult situations, I can pause without immediately reacting. |
| 18 | I remain curious about the nature of my experiences as they arise. |
| 19 | I am more invested in just watching my experiences as they arise, than in figuring out what they could mean. |
| 20 | I pay attention to sounds, such as clocks ticking, birds chirping, or cars passing. |
| 21 | My natural tendency is to put my experiences into words. |
| 22 | I rush through activities without being really attentive to them. |
| 23 | I approach my experiences by trying to accept them, no matter whether they are pleasant or unpleasant. |
| 24 | I am curious about my reactions to things. |
| 25 | I notice the smells and aromas of things. |
| 26 | I do jobs or tasks automatically without being aware of what I'm doing. |
| 27 | I am curious about what I might learn about myself by just taking notice of what my attention gets drawn to. |
| 28 | I think some of my emotions are bad or inappropriate and I shouldn't feel them. |
| 29 | I tend to react strongly to distressing thoughts and/or images. |

*Table 17 contd. ...*

*...Table 17 contd.*

| Characteristics in the First Iteration of the Mindfulness Heuristic |
|---|
| 30 When I have distressing thoughts or images, I judge myself as good or bad, depending what the thought/image is about. |
| 31 I have trouble noticing visual elements in art or nature, such as colors, shapes, textures, or patterns of light and shadow. |
| 32 I can usually describe how I feel at the moment in considerable detail. |
| 33 I am aware of my thoughts and feelings without over-identifying with them. |
| 34 I find myself doing things without paying attention. |
| 35 When I have distressing thoughts or images, I just notice them and let them. |

**Table 18:** Excerpt of CUNY mindfulness heuristic with rating scale.

| Mindfulness in Education Heuristic (Powietrzynska and Tobin 2015) |
|---|
| *For each characteristic, circle the numeral that best reflects your current state of mindfulness in this class. If necessary, provide contextual information that applies to your rating* |
| *5 = Very often or always; 4 = Often; 3 = Sometimes; 2 = Rarely; 1 = Hardly ever or never* |
| 1     I am curious about my emotions. |
| 2     I find words to describe my emotions. |
| 3     I allow thoughts to come and go without being distracted [or carried away] by them. |
| 4     I notice my emotions without reacting to them. |
| 5     I am kind to myself when things go wrong for me. |
| 6     I recover quickly when things go wrong for me. |
| 7     Even when I am focused, I use my senses to remain aware. |
| 8     When I am emotional, I notice [the pattern of] my breathing. |
| 9     When I am emotional, I notice my heart beat. |
| 10     I maintain a positive outlook. |
| 11     The way in which I express my emotions depends on what is happening. |
| 12     The way in which I express my emotions depends on who is present. |
| 13     I can focus my attention on learning. |
| 14     When I produce strong emotions, I can let them go. |
| 15     When my emotions change, I notice changes in my body (temperature). |
| 16     The way I position and move my body changes my emotions. |
| 17     I use breathing to [help] manage my emotions. |
| 18     I am kind to others when they are unsuccessful [or dealing with strong emotional states]. |
| 19     I can tell when something is bothering another person. |
| 20     I am aware of others' emotions from the tone of their voice [or body language]. |
| 21     I recognize others' emotions by looking at their faces. |
| 22     When I am with others, my emotions tend to become like their emotions. |

**Table 19:** Full version of mindfulness heuristic that included characteristics that address emotional styles, classroom dynamics and the ability to regulate emotions through physiological processes.

| Characteristics in the Mindfulness in Education Heuristic (Powietrzynska 2015) |
| --- |
| During this class: |

|    |    |
| --- | --- |
| 1 | I am curious about my feelings as they rise and fall. |
| 2 | I find words to describe the feelings I experience. |
| 3 | I identify distracting thoughts but let them go (without them influencing future action). |
| 4 | I am not hard on myself when I am unsuccessful. |
| 5 | I recover quickly when I am unsuccessful. |
| 6 | I pay attention to my moment-to-moment sensory experiences. |
| 7 | I am aware of the relationship between my emotions and breathing pattern. |
| 8 | I am aware of changes in my emotions and pulse rate. |
| 9 | I maintain a positive outlook. |
| 10 | I can tell when something is bothering the teacher. |
| 11 | I can tell when something is bothering other students. |
| 12 | The way in which I express my emotions depends on what is happening. |
| 13 | The way in which I express my emotions depends on who is present. |
| 14 | I can focus my attention on learning. |
| 15 | I feel compassion for myself when I am unsuccessful. |
| 16 | I feel compassion for others when they are unsuccessful. |
| 17 | When I produce strong emotions I easily let them go. |
| 18 | I gauge my emotions from changes in my body temperature. |
| 19 | I am aware of others' emotions from characteristics of their voices. |
| 20 | I am aware of my emotions being expressed in my voice. |
| 21 | I recognize others' emotions by looking at their faces. |
| 22 | I am aware of my emotions as they are reflected in my face. |
| 23 | My emotions are evident from the way I position and move my body. |
| 24 | The way I position and move my body changes my emotions. |
| 25 | I can tell others' emotions from the way they position and move their bodies. |
| 26 | I am aware of emotional climate and my role in it. |
| 27 | Seeking attention from others is not important to me. |
| 28 | Classroom interactions are characterized by winners and losers. |
| 29 | I meditate to manage my emotions. |
| 30 | I use breathing to manage my pulse rate. |
| 31 | I use breathing to manage my emotions. |

**Table 20:** Example of mindful listening and speech heuristic by the CUNY group.

| Compilation of Mindful Listening and Speech Heuristics (compilation (Tobin et al. 2015)) |
| --- |
| *When I participate in a conversation* |
| *Listening mindfully:* |

|    |    |
| --- | --- |
| 1 | I monitor the eyes of the speaker. |
| 2 | I show my respect for the speaker. |
| 3 | I express my opposition verbally and nonverbally to unethical speech. |
| 4 | While listening to others, my nonverbal actions project compassion and empathy to the speaker. |
| 5 | When a speaker says something with which I disagree, I try to learn from the difference. |
| 6 | I make sense of the speaker's facial expressions of emotion. |
| 7 | I make sense of the speaker's gestures. |
| 8 | I nod my head as a sign of attentiveness. |
| 9 | Following each utterance, I provide an appropriate pause to ensure that the speaking turn is finished. |
| 10 | When necessary, I seek clarification of the meaning of an utterance. |

*Table 20 contd. ...*

*...Table 20 contd.*

| Compilation of Mindful Listening and Speech Heuristics (compilation (Tobin et al. 2015)) |
|---|
| *When I participate in a conversation* |

*Listening mindfully:*

| | |
|---|---|
| 11 | When necessary, I request elaboration so as to expand the meaning of an utterance. |
| 12 | When necessary, I check my understanding of what has been said. |
| 13 | I ensure that my nonverbal actions do not breach the fluency of what is being said. |
| 14 | I use nonverbal actions to provide emotional synchrony with spoken text. |
| 15 | I ensure that my emotional response to spoken text does not stick and create difficulties in understanding subsequent utterances. |
| 16 | I listen for similarities and differences to what has been said previously. |
| 17 | I look for similarities and differences in the meanings represented verbally and nonverbally. |

*Mindful speech:*

| | |
|---|---|
| 1 | I act to balance the amount of time I talk. |
| 2 | When I have been speaking too long, I wind up my talking turn. |
| 3 | Before speaking, I pause to make sure the previous speaker has finished. |
| 4 | As I speak, I monitor others' emotions. |
| 5 | As I speak, I monitor my emotions. |
| 6 | When asynchronies occur, I try to understand them. |
| 7 | I try to make conversations with others successful. |
| 8 | When breaches in fluency occur, I try to repair them. |
| 9 | I do not increase the loudness of my voice to continue my talking turn. |
| 10 | I speak with a similar rhythm to previous speakers. |
| 11 | I maintain the focus of previous speakers. |
| 12 | I look for signs that others want to speak. |
| 13 | I am aware of how long I speak. |
| 14 | I create chances for others to speak. |
| 15 | I act to balance my speaking turns. |
| 16 | The loudness of my talk is appropriate. |
| 17 | I do not speak to hurt others. |
| 18 | My talk shows respect for others' perspectives. |
| 19 | I notice inequities in who has spoken and attempt make it more equitable by bringing into the discussion those left out. |

**Table 21:** Heuristic designed to for the discussion of 'thorny' topics in the classroom, which incorporates Davidson's six emotional styles principle.

| Excerpt from Alexakos et al.'s heuristic for discussion of 'thorny' issues[28] |
|---|
| *When discussing difficult social issues* |

| | |
|---|---|
| 2 | I am aware that my [speech] may have a hurtful effect on others, despite what my intentions may be. |
| 3 | I listen and consider changing my behavior if someone tells me that I am making them uncomfortable. |
| 4 | I am aware of my own prejudices and privileges, and critically reflect on my own habits, cultural practices, and how I create meaning. |
| 5 | Those feeling hurt by a discussion on thorny topics have the right to end the discussion when it becomes too much. |
| 6 | I try to create an environment that is inclusive, provides space for other voices, is mutually supportive, and is respectful to all. |
| 7 | I acknowledge deep emotional challenges without getting caught in them... |

*Table 21 contd. ...*

*...Table 21 contd.*

| Excerpt from Alexakos et al.'s heuristic for discussion of 'thorny' issues[28] |
|---|
| *When discussing difficult social issues* |

| | |
|---|---|
| 9 | When I have wronged or offended someone, I try to become aware of the transgression and any harm done, and if necessary, accept responsibility. |
| 18 | I acknowledge and respect views, values, knowledge systems, and life histories of others, though they may differ from my own. |
| 19 | I seek out and learn from opinions and experiences different than my own. |
| 20 | When others possess different identities, I use the opportunity to validate their experiences and perspectives.** |
| 21 | I try not to get disproportionately stuck when I am offended by opinions by others. |
| 22 | I forgive those who may offend me during respectful and well-meaning conversations related to thorny issues. |
| 23 | Even when I experience discomfort, my talk is engaging, intentional, meaningful, and honest. |
| 24 | Because it may hurt others, it is also mindful and kind. |
| 25 | I understand that I and others may not be able to use language to describe the emotions related to thorny issues. |
| 27 | I show myself compassion where appropriate. |
| 28 | My talk questions stereotypes and encourage acceptance of difference experiences.** |
| 29 | I try to involve those who do not seem to have been engaged.*** |
| 30 | I try to sustain difficult conversations to the point where authentic understanding and meaningful action occur, even when these conversations get uncomfortable.*** |
| 31 | I expect and accept a lack of closure when dealing with challenging issues.*** |
| | I trust my classmates to listen to my opinions and beliefs and I to theirs without thinking badly of one another. |

---

[28]   A number of these characteristics, marked by asterisk, were adapted by Alexakos et al. from other sources, such as: Bluestockings Bookstore's safer space policy (http:// bluestockings.com),* GLSEN's inclusive plans for LBBTQ students (http://www.glsen.org),** and Glenn Singleton's book *Courageous Conversations about Race*.***

# Appendix III—Breathing Exercises

The following are three examples of potential breathing exercises that could be used in the classroom.

**Figure 29:** Breathing exercise proposed to increase students' ability to observe thought processes, emotional states and bodily sensations.

**Mindful Breathing # 1**

This exercise involves simple observation of the breath, where the breath acts as an anchor for the students' awareness and attention. It can be used to deepen concentration while it also helps students let go of powerful, negative emotions and thoughts.

Procedure: Students breathe naturally and focus their attention on their breathing, either the rise and fall of their chest or they may focus on the sensation of the breath as it enters and leaves the nostrils. The teacher may guide this exercise with the following statements:

"Observe the natural pattern of your breathing, do not try to change it, just observe your natural inhalation and exhalation. If you become distracted by thoughts, judgements, sounds, feelings, emotions, etc. simpley let them go and return your attention to the rhythm of your breath."

Continue in this way for 3-5 min

**Mindful Breathing #2**

This exercise combines a mindful observation of the breath with an active manipulation of the breath. This exercise can be used not only to deepen concentration but also relax the body and mind.

Procedure: The students should begin by breathing naturally, then the teacher can instruct students to breath according to various patterns such as:
- Students breathe in for a count of three, pause for a count of three, then exhale for a count of three, pause for a count of three and then continue to breathe normally.
- Alternatively the cycles can be extended and varied, for instance, breathe in for a count of five, pause for a count of three, breath out for a count of seven, pause for a count of three. In general, this whole procedure can be repeated for a total of ten cycles.

Side note: You will not need to remind the students to let go of their thoughts as often during this exercise as the rhythmic counting normally keeps them concentrated on the task at hand. For maximum relaxation, try to have the outbreath extended longer than the inbreath.

**Figure 30:** Proposed breathing exercises for the classroom to encourage further emotional awareness as well as a means of calming the sympathetic system.

Note: It is very important to point out here that the goal of these type of breathing exercises cannot and should not be equated with depersonalization or disassociation, two psychological disorders where the person no longer feels like they are real but instead feel like they are an outside observer of their own thoughts or body. One technical difference is that depersonalization and disassociation periods are often longer and not entered into willingly. They are often subconscious coping mechanisms that have developed so that the person does not feel the full extent of their emotional trauma. What we are talking about here is very different. While individuals with depersonalization disorders report a loss of control over their thoughts, bodies and actions, the goal of the obtainment of the 'observer' or 'satellite' position is to gain *more* control. In essence, to take on the role of the guide. For this reason, it would possibly be better to refer to this conscious observer as the 'air traffic controller' to emphasize the type of observation and the goal of this observation. We are trying to support students in the ability to become aware of their thoughts so that they actually gain more control over their thought processes and their reactions.

# Appendix IV—Implementing Mindfulness

The most effective and ideal mode of implementing mindfulness practices would be a school-wide implementation that supports both students and teachers in the development of emotional awareness, emotional regulation and social intuition. According to research conducted by Nimrod Sheinman and Linor Hadar whole-school approaches appear to have the highest chance of success. This conclusion comes from research they conducted within the framework of *Israel's Whole School Mindfulness in Education* (WSMED), which has been implemented in various Israeli public primary schools (comprising 300–500 students each) since 2000. To offer the reader a better understanding of a whole-school approach to mindfulness interventions, a general overview of their program is reproduced below (Table 22).

Here we can see that such comprehensive programs require large-scale coordination from all educational levels and the involvement of the students' families. In the best-case scenario, these types of programs would be implemented nationwide (dare I say worldwide) at a primary school level to build a foundation of mindfulness that would assist students in their mental and psychological growth throughout their school experience. Many of the positive neurological effects of mindfulness practice are dependent upon the length of practice and thus the greatest potential for developing these beneficial mindsets would be through early childhood interventions. If such programs were offered in elementary schools or even middle school, high school biology teachers would then only need to reference these skills and then introduce the above heuristic as a specific intervention during lessons on evolution.

Despite the psychological and educational benefits of such proposed programs and the advantages it would offer to high school teachers, we are unfortunately still quite far away from a universal implementation of such programs. It is, nevertheless, possible for teachers to use components of these types of programs and apply them to certain classes for certain purposes, such as addressing the emotional intensity involved in classroom discussions surrounding human origins.

Another possibility is for teachers to directly introduce activities that support mindfulness in the form of breathing exercises (See Appendix III). Here biology

**Table 22:** Overview of the Major Components of Israel's Whole School Mindfulness in Education Program from Sheinman and Hadar 2016.

| Israel's Whole School Mindfulness in Education Program | |
|---|---|
| School's curriculum | ▪ Weekly mindfulness-based sessions, integral to each grade's curriculum, are offered to all classes on a weekly basis year after year.<br>▪ Sessions are taught by mindfulness facilitators, with each class's homeroom teacher present.<br>▪ Experiential learning is enhanced by inquiry, mindful-journaling and class discussions.<br>▪ Students are coached to apply mindfulness-based skills during school time as well as in real-life situation. |
| School's culture and/ or environment | ▪ A special room is designated by each school, without chairs and desks and with mattresses for each student.<br>▪ Each homeroom teacher is present in the weekly mindful sessions of their class.<br>▪ A monthly faculty meeting enhances teachers' gradual mindfulness-based teaching and learning.<br>▪ Mindfulness, led by teachers or children, is practiced before exams, at the beginning or end of a week, or when needed.<br>▪ Mindfulness is added to physical education classes, art classes, or as a short session at the beginning of a regular class.<br>▪ Mindfulness is being added to teachers' meetings. |
| Families and/or communities | ▪ Parents receive introductory information on the program.<br>▪ Parents receive updates on the mindfulness in schools project.<br>▪ Parents are invited to an introductory evening.<br>▪ Parents are invited to short mindfulness-based workshops for parents.<br>▪ Mindfulness is being practiced at school gatherings.<br>▪ Children are coached to introduce or teach mindfulness to their family members. |

teachers have an advantage as the concept of mindfulness can be integrated into science curricula fairly easily as much of mindfulness theory and practice is intrinsically linked to our physiology and can thus be easily linked to general lessons on anatomy and physiology. In the following table, I propose ways that mindfulness be introduced into the classroom in combination with lessons on anatomy, physiology or even evolution (Table 23).

So even though mindfulness is not a core component of central curricula, teachers do have the option to introduce components of mindfulness practices into the classroom in the form of heuristics or as part of other central lessons on anatomy and physiology. This is particularly true with the introduction of breathing exercises as these exercises can be used not only to teach the fundamentals of mindfulness but also of the intriguing connection between conscious breathing and activation of the parasympathetic nervous systems. Yet even if these are not used in the classroom but simply used by the teacher, it also changes the dynamic within the classroom. As the teacher becomes more aware of their own emotional state, whether it be fear, shame or annoyance, they too are able to better regulate how they choose to respond to a difficult encounter in the classroom.

**Table 23:** Integrating mindfulness practices into biology curriculum.

| Lessons Where Mindfulness Can Be Integrated into Science Content | |
|---|---|
| Anatomy and Physiology | <ul><li>Lessons on the brain and its anatomy</li><li>Lessons on neuroplasticity</li><li>Lessons on the nervous system, particularly about the nervous system(s)</li></ul> |
| General Health | <ul><li>Stress and its effect on the body, in particular a look at the link between stress and certain diseases and ways to counteract stress</li></ul> |
| Evolution | <ul><li>Evolution of stress-responses, threat-response, etc.</li><li>Evolution of various regions of the brain, in particular the difference between the neocortex and the limbic system</li></ul> |

# Appendix V—Specific Classroom Lessons

## Lesson 1: The Mystery of the Missing Chromosome

Lead-in: According to evolutionary theory, humans and great apes share a common ancestor. Yet chimpanzees (as well as gorillas and orangutans) have 48 chromosomes, while humans have only 46. In fact, all members of Hominidae have 24 pairs of chromosomes, just like modern great apes, with the exception of Homo Sapiens, Neanderthals, and Denisovans who have only 23 pairs of chromosomes. Where did this other pair of chromosomes go? Is it possible that this pair of chromosomes just got lost or deleted? The answer to that question is no, because a deletion of an entire pair of chromosomes would be lethal. So what else could have happened? Is it possible that maybe great apes and humans do not in fact share a common ancestor? Or is there another explanation for these missing chromosomes?

Procedure: Students should be given time to come up with possible explanation. Some students might use this opportunity to cast doubt on evolution and propose that humans and primates do not in fact share a common ancestor. Most students, though, will base their ideas upon their acceptance of evolution as a valid theory and come up with ways of explaining the deviation in the number of chromosomes, for instance that the common ancestor originally had 23 pairs and that one pair broke, creating an additional pair in subsequent hominids and great apes or that the original common ancestor had 24 pairs and that one of the pairs fused. If they are having trouble coming up with chromosomal change hypotheses, it may be necessary to help them along. Regardless of what types of ideas and hypotheses the students come up with, the teacher should guide their exploration of the topic, posing questions such as: How could that idea be tested? What sort of evidence would be found if your hypothesis is correct? What sort of evidence would disprove your hypothesis? These questions are particularly important for those students who propose that this deviation proves that great apes and humans do not share a common ancestor.

Once the students have come up with a hypothesis, it is important for them to come up with ways that they can test their ideas. Based on what they know about chromosomes, what would be the sign of chromosomes being fused? How could we test to see if chromosomes had split? The groups that proposed either a break of a fusion should come up with diagrams to explain their 'expected' findings. A sample diagram has been provided showing evidence of a fused chromosome (Figure 32).

Obviously, students will not be able to carry out real experiments but luckily, this chromosomal mystery has actually already been solved. Teachers can uncover

**Lesson: The Mystery of the Missing Chromosome**

Introduction: According to evolutionary theory, humans and great apes share a common ancestor. Yet chimpanzees (as well as gorillas and orangutans) have 48 chromosomes, while humans have only 46. In fact, all members of Hominidae also have 24 pairs of chromosomes, just like modern great apes. But Homo Sapiens, Neanderthals, and Denisovans have only 23 pairs of chromosomes. We know that it is not possible that it just got lost or deleted because a deletion of an entire pair of chromosomes would be lethal. So what else could have happened? Is it possible that maybe great apes and humans do not in fact share a common ancestor? Or is there another explanation for these missing chromosomes?

Procedure: Working in small groups students should come up with a hypothesis or hypotheses to explain the variation in the number of chromosomes. During this stage, teachers should give students total freedom and not become alarmed if some students propose that humans and primates do not, in fact, share a common ancestor. Teachers may guide students by telling them that they need to develop a testable hypothesis. So, for instance if we wanted to test the validity of evolution, we could hypothesize that the common ancestor originally had 23 pairs and that one pair broke, creating an additional pair in subsequent hominids or that the original common ancestor had 24 pairs and that one of the pairs fused in Homo Sapiens, Neanderthals and Denisovans.

Once the students have come up with a hypothesis, it is important for them to come up with ways that they can test their ideas. For example: Based on what they know about chromosomes, what would be a sign of chromosomes being fused or having been broken? (If necessary, teachers may need to review telomeres and centromeres) Teachers could also use or reproduce a simplified version of the 23 pairs of human chromosomes as seen on the next page.

Discussion: For the theory of evolution to be correct with regard to humans sharing a common ancestor with great apes, there must be some biological evidence to explain the variation in the number of chromosomes. This evidence is in fact present and has been well studied. Human chromosome 2 shows evidence of both vestigial telomeres and centromeres. As Hillier et al. published in Nature in 2005, "Human chromosome 2 is unique to the human lineage in being the product of a head-to-head fusion of two intermediate-sized ancestral chromosomes".

**Figure 31:** Printable overview of the lesson 'The Mystery of the Missing Chromosome'.

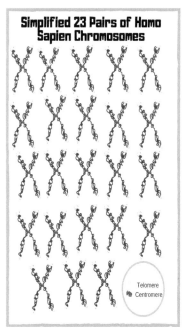

**Figure 32:** Example of a possible white-board image to be used for the lesson 'The Mystery of the Missing Chromosome'. Also possible as printable.

the solution in two possible ways. For a quick solution, teachers can use the diagram provided and give students time to 'analyze' the 23 pairs of chromosomes it in order see if there is evidence of either a break or a fusion, e.g., presence of additional centromeres. Using this diagram or a similar reproduction on a whiteboard, students should be able to quickly recognize the presence of vestigial telomeres and centromeres on human chromosome 2. Another means of unveiling the answer is to have the students conduct a literature search for the answer. This presents teachers also an additional opportunity to discuss reliable sources (peer-reviewed journals) versus unreliable sources (websites, particularly anti-scientific sites such as AnswersInGenesis). A search of the literature should lead students to this publication: *Generation and annotation of the DNA sequences of human chromosomes 2 and 4* in Nature (2005) by Hillier et al. For the teachers who chose to have the students analyze the diagram, they can follow-up the lesson by also reading excerpts from this publication.

Conclusion: Once students have the solution, it should be made very clear that this is how science works. The theory of evolution was proposed decades before scientists were even aware of chromosomes, meaning that the discovery and advanced understanding of genetics gave scientists new opportunities to test Darwin's theory. Instead of refuting the theory of evolution, advances in genetics has only substantiated the theory of evolution further. By the conclusion of the lesson the students should realize that this chromosomal deviation could have gone either way—either proving or disproving the theory of evolution. For evolution, particularly the common descent of modern apes and humans, to be correct there *must* be some biological evidence to explain the variation in the number of chromosomes and this is what in fact what was found as there is clear evidence of a chromosomal fusion as human chromosome 2 shows evidence of both vestigial telomeres and centromeres, as published in *Nature*

# Lesson: Testing the theory of evolution

Background: Many people falsely believe that it is not possible to test evolution because it is not possible to go back in time to observe the rise of new species, but this lesson shows how scientists can actually make and test hypotheses based on the theory of evolution. Before beginning this lesson: Make sure that the students are familiar with the scientific method, cladograms and geological records.

Introduction: Based on the theory of evolution all modern animals descended from more prehistoric animals. Based on this theory, there should be identifiable links between different types of animals, such as aquatic animals and terrestrial animals. Specifically, Darwin proposed that there should be fossils that illustrate the intermediate stage, for example, an animal such as Archaeopteryx which had the characteristics of both a bird and a reptile.

Goal: Students should become familiar with cladograms, phylogenetic trees, geological records and geological maps. They should learn how to use these data sources to test the theory of evolution by making and testing predictions about where intermediary fossils could be found.

Procedure: Working in groups, students should take on the role of researchers who are attempting to find the fossil remains of the 'transitional link' between fish and amphibians. Students should be provided with a cladogram or phylogenetic tree and a geological record so that they can make a prediction to about how long ago this intermediary animal would have lived. Depending on what information sources the students use, they should either come up with the name of a period, such as the Devonian period or with an approximate time frame, e.g. between 360 and 380 million years ago. Then using an interactive geological map, students should choose three potential digging sites where they would like to search for this intermediary animal.

Conclusion: Once students have finished, teachers can go over the story of how Daeschler, Shubin and Jenkins searched along the ancient Devonian shoreline in an attempt to do exactly this and how they ultimately discovered Tiktaalik.

**Figure 33:** Printable overview of the lesson 'Testing the Theory of Evolution'.

in 2005, "Human chromosome 2 is unique to the human lineage in being the product of a head-to-head fusion of two intermediate-sized ancestral chromosomes (Hillier et al. 2005)." If there is not time in during class for the entire exercise, then students could alternatively watch the original short presentation offered by Kenneth Miller.

## Lesson 2:  Testing the Theory of Evolution

Lead-in: The theory of evolution states that organisms have modified and changed very slowly over millions of years. This theory allows scientists to make the predication that if organisms changed very slowly over millions of years and that new species came into being and other species died out then it should be possible to find so-called 'missing links' or transitional fossils. The importance of these transitional fossils to the theory of evolution was recognized by Darwin as he developed his theory and stated that according to his theory, the fossil record should show "infinitely numerous transitional links" and yet this fossil evidence had not yet been found at the time that he developed his theory. This concerned Darwin as he stated, "Why then is not every geological formation and every stratum full of such intermediate links? Geology assuredly does not reveal any such finely graduated organic chain; and this, perhaps, is the gravest objection which can be urged against my theory (Darwin 1859; p. 280)."

After Darwin published his theory in 1859, paleontologists went out in search of these fossils to test Darwin's new theory and to see if there was any fossil evidence that could explain the origin of birds. By 1860, a fossilized feather had been found in Germany in limestone dating back to the late Jurassic period. This feather was subsequently described as *Archaeopteryx lithographica* by Christian Erich Hermann von Meyer in 1861. Just two years later, Richard Owen described a nearly complete skeleton of *Archaeopteryx lithographica*, which included many reptilian features, such as clawed forelimbs and a bony tail, despite its overall similarity in appearance to a bird (1863). Thomas Henry Huxley recognized that Archaeopteryx was a transitional organism that possessed characteristics typical of both birds and reptiles and that it counted as one of Darwin's proposed 'intermediate links' based on the fact that it lived 155–150 million years ago and possessed both the characteristics of a bird and a reptile as he stated, "I think I have shown cause for the assertion that the facts of paleontology, so far as birds and reptiles are concerned, are not opposed to the doctrine of evolution, but, on the contrary, are quite such as that doctrine would lead us to expect; for they enable us to form a conception of the manner in which birds may have been evolved from reptiles, and thereby justify us in maintaining the superiority of the hypothesis that birds have been so originated to all hypotheses which are devoid of an equivalent basis of fact (1868, p. 75)." By 1870, the iconic 'Berlin specimen' of *Archaeopteryx*, had been discovered and provided further evidence of its role as a transitional fossil due to the presence of a complete set of reptilian teeth. This rapid succession of discoveries that quickly substantiated Darwin's theory may seem serendipitous to some, but they are actually examples of how the theory of evolution can be tested using the hypothetico-deductive method. To test the theory of evolution, we first need to establish three points: (1) *What?* What transition are we interested in, i.e., aquatic animals to land animals, or land to air transition? (2) *When?* When do we believe that this transition should have occurred? and (3) *Where?* Where would this age of rock be present on the surface of the earth?

Procedure: The story of the discovery of *Archaeopteryx* can be used as an introduction to the idea of testing evolution using the hypothetico-deductive method. Students should then use the information available to them to come up with a hypothesis of where one could look for transitional fossils from water animals to amphibians. In this exercise, question one has already been answered but students should come up with the answers for (2) *When?* and (3) *Where?* In other words, according to what we know about evolution, when do we believe that water animals transitioned to land and where on earth do we have access to this age of stone to search for these hypothetical fossil remains?

Using phylogenetic trees and geological time scales, students should be able to come up the Devonian period as the logical time for the transition to have occurred (Figure 34). Then using an interactive geologic map of the world (such as the one found at https://macrostrat.org/map), which shows the various age of surface rock, students should come up with a potential area where they could search for Devonian fossils (Figure 25).

Like in the previous lesson, students are again unable to conduct the actual research but once students have come up with their hypothesized time of transition and place where they could dig for fossils, teachers can go over the story of a research group who actually did conduct this exact experiment: In 1999, Edward Daeschler, Neil Shubin and Farish Jenkins began to actively search for a particular fossil based on evolutionary and geological data. Using the hypothetico-deductive method, Daeschler and his team made a prediction based on the evolutionary theory. If fish gave rise to land animals then there should be evidence of an organism that shows both fish and amphibian characteristics and according to evolutionary theory, such an organism should have lived between 360 and 380 million years ago. So, it should be possible to analyze rock from this period of time and find evidence of these transitional organisms. In order to test their hypothesis, these scientists travelled to the Canadian arctic where large amounts of stone from an ancient Devonian shore line are exposed and there, they began to search for this fossil in order to test their prediction (and ultimately the theory of evolution).

In 2004, their efforts were rewarded with the discovery of what would later be called Tiktaalik on Ellesmere Island in Nunavut, Canada. According to Shubin, their expedition illustrates how the scientific method can be applied to evolutionary science and that because Tiktaalik has both the characteristics of primitive fish and amphibians it fills in the gap between water dwellers and land dwelling tetrapods (2009).

Conclusion: Once students have the solution, it should be made very clear that the work of Daeschler, Shubin and Jenkins shows not only the validity of the theory of evolution but how it can be tested using the scientific method. Not only are the transitional fossils present, but they can be actively searched for using the data we have from phylogenetic trees. More information on their work can be found by reading their 2006 publication in Nature, "A Devonian tetrapod-like fish and the evolution of the tetrapod body plan."

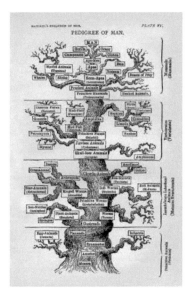

1) Use cladograms or phylogenetic trees to determine what transition to examine. e.g. from fish to amphibians.

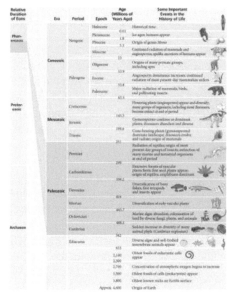

2) Use geological time scales to decide when the transition most likely occurred.

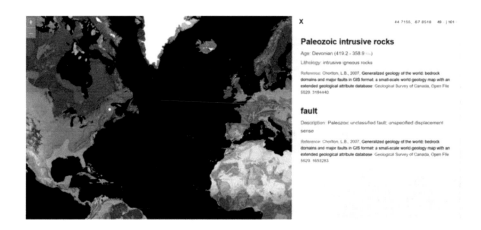

3) Use interactive geological maps to decide where we could dig for fossils to test our hypothesis.

**Figure 34:** Examples of tools to be used during the lesson 'Testing the Theory of Evolution'.

# Lesson: Teaching Common Humanity

In theory, students can actively develop a sense of common humanity by sitting together with other students and by going through a series of specific questions that is designed to identify how different backgrounds lead to different viewpoints and to help participants identify what they have in common regardless of their differences in opinion. The ultimate goal of this exercise is to realize how we are all influenced by our upbringing, culture, religious beliefs and personal experiences. By realizing how our worldviews are influenced through external factors, we can better understand that these viewpoints do not define us as a human being and that each of us is much greater than a given set of opinions or beliefs. The second goal of this exercise is to also realize that regardless of whether we have very different belief systems than another person, we may still share a great deal in common because regardless of whether someone is a democrat or a republican, a creationist or a new atheist, we often share common interests or concerns. It might be as simple as a love for dogs or concern for a sick relative, the point of this exercise is to encourage a conversation that allows seemingly very different individuals to identify what unites them.

**Table 24:** Proposed classroom exercise to decrease our natural tendency to 'other'.

| Understanding Our Differences & Identifying Our Common Humanity | |
| --- | --- |
| **How are we different?** | **What do we have in common?** |
| Upbringing | Upbringing |
| Culture | Culture |
| Personal History | Personal History |
| Education | Education |
| Religious Upbringing | Religious Upbringing |
| Hobbies/Interests | Hobbies/Interests |
| Traditions | Traditions |
| Likes/Dislikes | Likes/Dislikes |
| Etc. | Etc. |
| **What do we disagree about?** | **What can we agree about?** |
| Values | Values |
| Political Views | Political Views |
| Worldviews | Worldviews |
| Belief Systems | Belief Systems |
| Fears | Fears |
| Hopes | Hopes |
| Plans | Plans |
| Entertainment | Entertainment |
| Etc. | Etc. |
| **Follow up questions:** | |
| What do you and your partner have in common? What are the biggest differences between you and your partner? What do the two of you agree about? What are your biggest disagreements? | |
| **Conclusion/Reflection** | |
| Now that you understand how you and your partner are different from one another—can you identify reasons (e.g., from your different upbringings, cultures, personal histories, etc.) that would allow you to better understand why you and your partner disagree on certain points? | |

Specific procedure: Students should first discuss the questions in part I in the form of a conversation or interview. Once they have gone through the questions in the first part of the exercise, students should then engage in a period of reflection regarding the main ways in which they differ from their conversation partner and what points they share in common. Then looking back on what they know about their conversation partner's upbringing, culture, etc., they can better understand why they might have varying positions, viewpoints, beliefs, etc. This reflection period should allow students to better recognize how we are all influenced by the social context in which we are raised. Understanding this enables us to pause and stop judging one another based on the pretext of our own lives and experiences. After the conclusion of the exercise, students should share with the class what they found out that they have in common, e.g., we are both Yankee fans, we are both worried about getting into college, etc.

# References

Alexakos, K., L.D. Pride, A. Amat, P. Tsetsakos, K.J. Lee, C. Paylor-Smith et al. 2016. Mindfulness and discussing "thorny" issues in the classroom. Cultural Studies of Science Education, 11: 741–769. doi:10.1007/s11422-015-9718-0.

Alsop, S. and M. Watts. 2003. Science education and affect. International Journal of Science Education, 25: 1043–1047.

American Academy of Child and Adolescent Psychiatry. 2016. Teen Brain: Behavior, Problem Solving, and Decision Making. https://www.aacap.org/AACAP/Families_and_Youth/Facts_for_Families/FFF-Guide/The-Teen-Brain-Behavior-Problem-Solving-and-Decision-Making-095.aspx, AACAP.

Baer, R.A., E. Walsh and E.L.B. Lykins. 2009. Assessment of Mindfulness. In: D.F. (Ed.). Clinical Handbook of Mindfulness. New York, NY: Springer.

Balmer, R. 2012. Of God and Caesar. God in America Series, PBS.

Barbour, H. and J.W. Frost. 1998. The Quakers. Greenwood Press.

Barker, G. (Writer) and D. Belton (Director). 2012. Of God and Ceasar. In: Barker, G., J. Powell, D. Espar, D. Fanning and M. Samels (Producer), God in America. Frontline: PBS.

Becker, E. 1973. The Denial of Death. New York, Free Press Paperbacks.

Behar-Horenstein, L.S. 2006. Motivation. pp. 308–314. In: Feinstein, S. (Ed.). The Praeger Handbook of Learning and the Brain (Vol. 2): Greenwood Publishing Group.

Behe, M.J. 1996. Darwin's Black Box: The Biochemical Challenge to Evolution New York, NY, The Free Press.

Belton, D. (Writer). 2012a. A New Adam, God in America: PBS.

Belton, D. (Writer). 2012b. New Eden, God in America: PBS.

Berger, K.S. 2008. The developing person through the life span, Worth.

BGEA. 2018. Billy Graham: Pastor to Presidents. Retrieved from https://billygraham.org/story/billy-graham-pastor-to-presidents-2/.

Bishop, S.R., M. Lau, S. Shapiro, L. Carlson, N.D. Anderson, J. Carmody et al. 2004. Mindfulness: a proposed operational definition. Clin. Psychol. Sci. Pract., 11: 230–241.

Blakemore, S.-J. 2014. Adolescent brain development. Retrieved from https://thinkneuroscience.wordpress.com/2014/01/22/adolescent-brain-development/.

Blancke, S., H.H. Hjermitslev and P.C. Kjærgaard. 2014. Creationism in Europe Baltimore, MA, John Hopkins UP.

Bourdieu, P. and L.J. Wacquant. 1992. An Invitation to Reflexive Sociology. Chicago, The University of Chicago Press.

Buss, D.M. 1995. Evolutionary psychology: A new paradigm for psychological science. Psychological Inquiry, 6: 1–30.

Buss, D.M. 1998. The Handbook of Evolutionary Psychology. Hoboken, Wiley.

Cartwright, R.A. 2004. The Biology Teacher Next Door: Susan Epperson at Evolution. The Panda's Thumb. Retrieved from https://pandasthumb.org/archives/2004/07/the-biology-tea.html.

Chang, M.L. 2009. An appraisal perspective of teacher burnout: Examining the emotional work of teachers. Educational Psychology Review, 21L: 193–218.

CMind. 2015. Contemplative Practices.

Cohen, A.D. 2006. Social, emotional, ethical, and academic education: Creating a climate for learning, participation in democracy, and well-being. Harvard Educational Review, 76: 201–237.

Colt, S. and T. Jennings (Writers) and D. Belton (Director). 2012. A New Light, God in America: PBS.

Curry, A. 2008. Gobekli Tepe: The World's First Temple? Predating Stonehenge by 6,000 years, Turkey's stunning Gobekli Tepe upends the conventional view of the rise of civilization. Smithsonian Magazine.

Daeschler, E.B., N.H. Shubin and F.A. Jenkins. 2006. A Devonian tetrapod-like fish and the evolution of the tetrapod body plan. Nature, 440: 757–763. doi:10.1038/nature04639.

Darwin, C. 1859. On the Origin of Species. London, John Murray.

Davidson, R.J. and S. Begley. 2012. The emotional life of your brain: how its unique patterns affect the way you think, feel, and live—and how you can change them. New York, Hudson Street Press.

Davidson, R.J., J. Dunne, J.S. Eccles, A. Engle, M. Greenberg, P. Jennings et al. 2012. Contemplative practices and mental training. Child Development Perspectives, 6: 146–153.

Demastes-Southerland, S., R. Good and P. Peebles. 1995. Student's conceptual ecologies and the process of conceptual change in evolution. Science Education, 79: 637–666.

Diner, H.R. 2012. A New Light. God in America, PBS.

Dixon, T. 2008. Science and Religion: A Very Short Introduction. Oxford, Oxford University Press.

Dobrzycki, J. and L. Hajdukiewicz. 1969. Micholi Kopernik, Polish Biographical Dictionary (Vol. XIV). Wrocław, Polish Academy of Sciences.

Dobzhansky, T. 1973. Nothing in Biology Makes Sense Except in the Light of Evolution. The American Biology Teacher, 35: 125.

Dole, J.A. and G.M. Sinatra. 1998. Reconceptualizing change in the cognitive construction of knowledge. Educational Psychologist, 33: 109–128.

Duit, R. 1999. Conceptual change approaches in science education. pp. 263–282. *In*: Schnotz, W., S. Vosniadou and M. Carretero (Eds.). New Perspectives on Conceptual Change. Kidlington, Oxford: Elsevier Science.

Duit, R. and D. Treagust. 2003. Conceptual change: a powerful framework for improving science teaching and learning. International Journal of Science Education, 25: 671–688.

Eisenstein, E.L. 1980. The Printing Press as an Agent of Change. Cambridge, Cambridge University Press.

Ellison, C.G., A.M. Burdette and T.D. Hill. 2009. Blessed assurance: Religion, anxiety, and tranquility among US adults. Social Science Research, 38: 656–667.

Felver, J.C. and P.A. Jennings. 2016. Applications of mindfulnessbased interventions in school settings: an introduction. Mindfulness, 7: 1–4.

Flank, L. 2007. Deception by Design: The Intelligent Design Movement in America. St. Petersburg, FL, Red and Black Publishers.

Forrest, B. and P.R. Gross. 2007. Creationism's Trojan Horse: The Wedge of Intelligent Design. New York, NY, Oxford University Press.

Fronsdal, G. 1998. Insight meditation in the United States: Life, liberty, and the pursuit of happiness. pp. 163–182. *In*: Prebish, C.S. and K.K. Tanaka (Eds.). The Faces of Buddhism in America: University of California Press.

Fronsdal, G. 2008. The Issue At Hand: Essays On Buddhist Mindfulness Practice Paperback, Bookland.

Gable, L.S., T.J. Reis and J.A. Elliot. 2000. Behavioral activation and inhibition in everyday life. Journal of Personality and Social Psychology, 78: 1135–1149.

Gibbs, C. 2003. Explaining effective teaching: Self-efficacy and thought control of action. Journal ofEducational Enquiry, 4.

Gordin, M.D. 2012. The PseudoScience Wars: Immanuel Velikovsky and the Birth of the Modern Fringe. Chicago, University of Chicago Press.

Gotink, R.A., P. Chu, J.J.V. Busschbach, H. Benson, G.L. Fricchione and M.G.M. Hunink. 2015. Standardised mindfulness-based interventions in healthcare: an overview of systematic reviews and metaanalyses of RCTs. PLoS One, 10: e0124344.

Gould, S.J. 1997. Nonoverlapping magisteria. Natural History, 106: 16–22.

Grant, E. 1984. Science in the medieval university. *In*: James M. Kittleson and Pamela J. Transue (Eds.). Rebirth, Reform and Resilience: Universities in Transition, 1300–1700 (pp. 68). Columbus: Ohio State University Press.

Greenberg, M.T., R.P. Weissberg, M.U. O'Brien, J.E. Zins, L. Fredericks and H. Resnik. 2003a. Enhancing school-based prevention and youth development through coordinated social, emotional, and academic learning. American Psychologist, 58: 466–474.

Greenberg, M.T., R.P. Weissberg, M.U. O'Brien, J.E. Zins, L. Fredericks, H. Resnik et al. 2003b. Enhancing school-based prevention and youth development through coordinated social, emotional, and academic learning. American Psychologist, 58: 466–474.

Greenberg, M.T. and A.R. Harris. 2012. Nurturing Mindfulness in Children and Youth: Current State of Research. Child Development Perspectives, 6: 161–166.

Gregoire, M. 2003. Is it a challenge or a threat? A dual-process model of teachers' cognition and appraisal processes during conceptual change. Educational Psychology Review, 15: 147–179. doi:10.1023/A:1023477131081.

Gribben, C. 2011. Evangelical Millennialism in the Trans-Atlantic World, 1500–2000. New York, Palgrave McMillan.

Ham, K. 2012. The Lie: Evolution (revised and expanded). Green Forest, AR, Master Books, Inc.

Ham, K. 2013. Six Days: The Age of the Earth and the Decline of the Church. Green Forest, AR, Master Books, Inc.

Hanson, R. 2011. Paper Tiger Paranoia, Spirit Rock.

Hanson, R. 2013. Hardwiring Happiness: The New Brain Science of Contentment, Calm, and Confidence. New York, Harmony Books.

Harari, Y.N. 2014. Sapiens: A Brief History of Humankind Harvill Secker.

Hart, T. 2004. Opening the contemplative mind in the classroom. Journal of Transformative Education, 2: 28–46.

Haviland, W.A., H.E.L. Prins, D. Walrath and B. McBride. 2010. Anthropology: The Human Challenge, Wadsworth.

Hemminger, H. 2009. Und Gott schuf Darwins Welt—Der Streit um Kreationismus, Evolution und Intelligentes Design. Gießen, Brunnen Verlag.

Hen, M. and A. Sharabi-Nov. 2014. Teaching the teachers: emotional intelligence training for teachers. Teaching Education, 25: 375–390. doi:10.1080/10476210.2014.908838.

Hillier, L.W., T.A. Graves, R.S. Fulton, L.A. Fulton, K.H. Pepin, P. Minx et al. 2005. Generation and annotation of the DNA sequences of human chromosomes 2 and 4. Nature, 434: 724. doi:10.1038/nature03466. https://www.nature.com/articles/nature03466#supplementary-information.

Holzel, B.K., J. Carmody, M. Vangel, C. Congleton, S.M. Yerramsetti, T. Gard et al. 2011. Mindfulness practice leads to increases in regional brain gray matter density. Psychiatry Res., 191: 36–43. doi:10.1016/j.pscychresns.2010.08.006.

Hood, R.W., P.C. Hill and W.P. Williamson. 2005. The Psychology of Religious Fundamentalism. New York, Guilford Press.

Hopper, S. 2006. The rise of the 'New Atheists'. Retrieved from http://edition.cnn.com/2006/WORLD/europe/11/08/atheism.feature/index.html.

Horstman, B.M. 2002. BILLY GRAHAM: A MAN WITH A MISSION (SPECIAL SECTION). The Cincinnati Post. Retrieved from https://www.highbeam.com/doc/1G1-87912863.html.

Hovland, C.I., I.L. Janis and H.H. Kelley. 1953. Communication and Persuasion. New Haven, Yale University Press.

Huff, T.E. 2017. The Rise of Early Modern Science Islam, China, and the West (3rd ed.). Massachusetts, Harvard University.

Humes, E. 2007. Monkey Girl. New York, HarperCollins Publisher.

Hummer, R.A., R.G. Rogers, C.B. Nam and C.G. Ellison. 1999. Religious Involvement and U.S. Adult Mortality. Demography, 36: 273. doi:10.2307/2648114.

Huxley, T.H. 1868. On the animals which are most nearly intermediate between birds and reptiles. Annals & Magazine of Natural History, 2: 66–75.

Hyers, C. 1984. The Meaning of Creation: Genesis and Modern Science, Westminster John Knox Press.

Immordino-Yang, M. and A. Damasio. 2007. We feel, therefore we learn: The relevance of affective and social neuroscience to education. Mind, Brain and Education, 1: 3–10.

Inzlicht, M., A.M. Tullett and M. Good. 2011. The need to believe: a neuroscience account of religion as a motivated process. Religion, Brain & Behavior, 1: 192–212. doi:10.1080/2153599x.2011.647849.

Ireland, T. 2014. What Does Mindfulness Meditation Do to Your Brain? Scientific America.

Jennings, P.A. 2007. Cultivating emotional balance in the classroom. Paper presented at the American Psychological Association Annual Convention, San Francisco.

Jennings, P.A. and M.T. Greenberg. 2011. The Prosocial Classroom: Teacher Social and Emotional Competence in Relation to Student and Classroom Outcomes. Review of Educational Research, 79: 491–525. doi:10.3102/0034654308325693.

Kazin, M. 2012. A New Light. God in America, PBS.

Kemeny, M.E., C. Foltz, M. Cullen, P. Jennings, O. Gillath, B.A. Wallace et al. 2012. Contemplative/ Emotion Training Reduces Negative Emotional Behavior and Promotes Prosocial Responses. Emotion, 12: 338–350. doi:10.1037/a0026118.supp.

Khalturin, K. and T. Bosch. 2007. Self/nonself discrimination at the basis of chordate evolution: limits on molecular conservation. Current Opinion in Immunology, 19: 4–9.

Khoury, B., M. Sharma, S.E. Rush and C. Fournier. 2015. Mindfulness-based stress reduction for healthy individuals: a metaanalysis. Journal of Psychosomatic Research, 78: 519–528.

Kidd, T. 2007. The Great Awakening: The Roots of Evangelical Christianity in Colonial America, Yale University.

Kindt, J. and T. Latty. 2018. Guide to the classics: Darwin's On the Origin of Species Read more at: Phys. Org. Retrieved from https://phys.org/news/2018-05-classics-darwin-species.html#jCp.

King, R.E. 1997. When worlds collide: politics, religion, and media at the 1970 East Tennessee Billy Graham Crusade. Journal of Church and State.

Koenig, H.G., J.C. Hays, D.B. Larson, L.K. George, H.J. Cohen, M.E. McCullough et al. 1999. Does Religious Attendance Prolong Survival? A Six-Year Follow-Up Study of 3,968 Older Adults. The Journals of Gerontology Series A: Biological Sciences and Medical Sciences, 54: M370–M376. doi:10.1093/gerona/54.7.M370.

Kremenitzer, J.P. and R. Miller. 2008. Are you a highly qualified, emotionally intelligent early childhood educator? Young Children, 63: 106–112.

Kron, P. 2018. Conversations on Creation. *In*: Neanderthals in the Classroom by Elizabeth Watts. 2019. Taylor and Francis.

Kübler-Ross, E. 1969. On Death & Dying, Simon & Schuster/Touchstone.

Kuyken, W., K. Weare, O.C. Ukoumunne, Rachael Vicary, N. Motton, R. Burnett et al. 2013. Effectiveness of the Mindfulness in Schools Programme: Non-randomised Controlled Feasibility Study. The British Journal of Psychiatry, 203: 126–131.

Landau, M.J., S. Solomon, T. Pyszczynski and J. Greenberg. 2007. On the compatibility of terror management theory and perspectives on human evolution. Evolutionary Psychology, 5: 476–519.

Larson, E.J. 1997. Summer for the Gods: The Scopes Trial and America's Continuing Debate Over Science and Religion, Basic Books.

Larson, E.J. 2012. A New Light. God in America PBS.

Lattimer, R.P. 2013. Exhibit "B". Response of Citizens for Objective Public Education, Inc. (COPE) to the January 2013 Draft of National Science Education Standards (the Standards) and the Framework for K-12 Science Education (the Framework) upon which the Standards are based. Original Complaint in COPE et al. v. Kansas SBOE et al., Case 5:13-cv-04119-KHV-JPO.

Laukenmann, M., M. Bleicher, S. Fuß, M. Gläser-Zikuda, P. Mayring and C.v. Phöneck. 2003. An investigation of the influence of emotional factors on learning in physics instruction. International Journal of Science Education, 25: 489–507.

Lehrer, P.M. and R. Gevirtz. 2014. Heart rate variability biofeedback: how and why does it work? Frontiers in Psychology, 5. doi:10.3389/fpsyg.2014.00756.

Lindberg, D.C. 2007. The Beginnings of Western Science. Chicago, IL, University of Chicago Press.

Linehan, M. 1993. Cognitive-Behavioral Therapy of Borderline Personality Disorder. New York, The Guilford Press.

Linnenbrink, E.A. and P.R. Pintrich. 2004. Role of affect in cognitive processing in academic contexts. pp. 57–87. *In*: Dai, D.Y. (Ed.). Motivation, Emotion, and Cognition: Integrative Perspectives on Intellectual Functioning and Development. Mahwah, NJ: Lawrence Erlbaum Associates.

Longfield, B.J. 2000. For Church and Country: The Fundamentalist-Modernist Conflict in the Presbyterian Church. The Journal of Presbyterian History, 78: 35–50.

Lowel, S. and W. Singer. 1992. Selection of intrinsic horizontal connections in the visual cortex by correlated neuronal activity. Science 255: 209–212.

Lugo, L., S. Stencel, J. Green, G. Smith, D. Cox, A. Pond et al. 2008. U.S. Religious Landscape Survey— Religious Beliefs and Practices: Diverse and Politically Relevant. Retrieved from Washington, D.C.

Luhrmann, T.M. 2013. The Benefits Of Church. New York Times, p. SR9.

Lyerly, C.L. 2012. A New Eden. God in America.

MacLean, P.D. 1990. The triune brain in evolution: role in paleocerebral functions. New York, Plenum Press.

Marris, P. 1986. Loss and Change. London, Routledge.

Masuch, A., P. Perialis and H.-J. Lehnert. 2007. Die Pflanze trinkt. Wie man Schülervorstellungen zu botanischen Themen herausfinden und unterrichtlich nutzen kann. Sache, Wort, Zahl, 35: 52–57.

Matzke, N.J. 2010. The Evolution of Creationist Movements. Evolution: Education and Outreach, 3: 145–162. doi:10.1007/s12052-010-0233-1.

McGrath, A.E. 1998. Historical Theology, An Introduction to the History of Christian Thought. Oxford, Blackwell Publishers.

Mencken, H.L. 1925. The Tennessee Circus. Baltimore Evening Sun.

Mendelson, T., M.T. Greenberg, J.K. Dariotis, L.F. Gould, B.L. Rhoades and P.J. Leaf. 2010. Feasibility and preliminary outcomes of a school-based mindfulness intervention for urban youth. Journal of Abnormal Child Psychology, 38: 985–994.

Miller, J.C. 1966. The First Frontier: Life in Colonial America, University Press of America.

Miller, J.D., E.C. Scott and S. Okamoto. 2006. Science communication. Public acceptance of evolution. Science, 313: 765–766. doi:10.1126/science.1126746.

Miller, R. 2012. A New Light. God in America.

Morris, H.M. 1961. The Genesis Flood, Presbyterian and Reformed Publishing Company.

Morris, H.M. 1972. The Remarkable Birth of Planet Earth, Bethany House Publications.

Morris, H.M. 1974. Many Infallible Truths. Green Forest, Master Books, Inc.

Morris, H.M. 1989. The Long War Against God: the history and impact of the creation/evolution conflict. Green Forest, AR, Master Books, Inc.

Moustakas, C. 1990. Heuristic research: design, methodology, and applications. Newbury Park, CA, SAGE Publications Inc.

Napoli, M., P.R. Krech and L.C. Holley. 2005. Mindfulness Training for Elementary School Students: The Attention Academy. Journal of Applied School Psychology, 21: 99–125.

Newport, F. 2018. In the News: Billy Graham on 'Most Admired' List 61 Times. Gallup.

Noddings, N. 2005. The Challenge to Care in Schools: An Alternative Approach to Education. New York, Teachers College, Columbia University.

Numbers, R. 1992. The Creationists. Berkeley and LA, University of California Press.

Numbers, R. 2014. Foreword. *In*: Blancke, S. (Ed.). Creationism in Europe. Baltimore: Johns Hopkins UP.

Numbers, R.L. 1998. Darwinism Comes to America, Harvard University Press.

Numbers, R.L. 2006. The Creationists: From Scientific Creationism to Intelligent Design, Expanded Edition, Harvard University Press.

O'Donohue, J. 2009. Eternal Echoes: Celtic Reflections on Our Yearning to Belong, Harper Collins.

Ogilvie, D.M. 2010. Soul Beliefs: Causes and Consequences. Retrieved 14 October 2018.

Orr, J., B.B. Warfield and C. Morgan (Eds.). 1910. The Fundamentals: A Testimony to the Truth (Vol. 1). Chicago: Testimony Publishing Company.

Owen, R. 1863. On the *Archeopteryx* [sp.] of von Meyer, with a description of the fossil remains of a long-tailed species, from the lithographic stone of *Solenhofen* [sp.]. Philosophical Transactions of the Royal Society of London, 153: 33–47.

Pace, E.J. 1922. The Descent of the Modernists. *In*: Seven Questions in Dispute by William Jennings Bryan, 1924, New York: Fleming H. Revell Company.

Padian, K. 2010. How to Win the Evolution War: Teach Macroevolution! Evolution: Education and Outreach, 3: 206–214. doi:10.1007/s12052-010-0213-5.

Paley, W. 1802. Natural Theology or Evidences of the Existence and Attributes of the Deity. Philadelphia, John Morgan.

Pannekoek, J.N., S.J.v.d. Werff, P.H. Means, B.G.v.d. Bulk, D.D. Jolles, I.M. Veer et al. 2014. Aberrant resting-state functional connectivity in limbic and salience networks in treatment--naïve clinically depressed adolescents. J. Child Psychol. Psychiatry, 55: 1317–1327.

Pekrun, R. and E.J. Stephens. 2012. Academic emotions. pp. 3–31. *In*: Harris, K.R., S. Graham, T. Urdan, S. Graham, J.M. Royer and M. Zeidner (Eds.). Apa Educational Psychology Handbook, vol 2: Individual Differences and Cultural and Contextual Factors. Washington, DC: American Psychology Association.

Pelham, B. 2009. Awareness, Opinions About Global Warming Vary Worldwide. from Gallup.

Perry, B.D. and M. Szalavitz. 2011. Born for Love: Why Empathy Is Essential—and Endangered, William Morrow Paperbacks.

Pettegree, A. 2000. The Reformation World, Routledge.

PEW Research Center. 2008. The Religious Composition of the United States U.S. Religious Landscape Survey: Religious Affiliation. http://www.pewforum.org/2008/02/01/chapter-1-the-religious-composition-of-the-united-states/.

Piaget, J. 1985. Equilibration of Cognitive Structures: The Central Problem of Intellectual Development. Chicago, University of Chicago Press.

Powietrzynska, M., K. Tobin and K. Alexakos. 2014. Facing the grand challenges through heuristics and mindfulness. Cultural Studies of Science Education, 10: 65–81. doi:10.1007/s11422-014-9588-x.

Powietrzynska, M. 2015. Heuristics for mindfulness in education and beyond. pp. 59–80. *In*: Milne, C., K. Tobin and D. DeGennaro (Eds.). Sociocultural Studies and Implications for Science Education (Vol. 12). Springer, Dordrecht: Cultural Studies of Science Education.

Powietrzynska, M. and K. Tobin. 2015. Mindfulness and science education. pp. 1–7. *In:* R. Gunstone (Ed.). Encyclopedia of Science Education: Springer Netherlands.

Prothero, S. 2012a. A New Adam. God in America Series, PBS.

Prothero, S. 2012b. Of God and Caesar. God in America Series, PBS.

Racevska, E. 2018. Evolutionary psychology. pp. 1–14. *In*: Vonk, J. and T.K. Shackelford (Eds.). Encyclopedia of Animal Cognition and Behavior: Springer International Publishing AG.

Ramakrishnan, J. 2013. Brain Based Learning Strategies. International Journal of Innovative Research and Studies, 2: 236–242.

Ratzinger, J. 1995. In the Beginning: A Catholic Understanding of the Story of Creation and the Fall, Eerdmans.

Reiss, M.J. 2008. Teaching evolution in a creationist environment: an approach based on worldviews, not misconceptions. The School Science Review, 90: 49–56.

Richardson, B.G. and M.J. Shupe. 2003. The importance of teacher self-awareness in working with students with emotional and behavioral disorders. Teaching Exceptional Children, 36: 8–13.

Ridder-Symoens, H.d. (Ed.). 1992. A History of the University in Europe (Vol. I: Universities in the Middle Ages): Cambridge University Press.

Rosenau, J. 2015. Evolution, the Environment, and Religion.

Ruse, M. 2001. The Evolution Wars: A Guide to the Debates. New Brunswick, NJ, Rutgers University Press.

Salomon, G. and T. Globerson. 1987. Skill may not be enough: The role of mindfulness in learning and transfer. International Journal of Education Research, 11: 623–637.

Santorum, R. 2011. The Rick Santorum Interview/*Interviewer: R. Limbaugh*. Rush Limbaugh Show, https://www.rushlimbaugh.com/daily/2011/06/08/the_rick_santorum_interview/.

Schonert-Reichl, K.A. and R.W. Roeser. 2016. Handbook of Mindfulness in Education, Springer.

Schutz, P.A., J.Y. Hong, D.I. Cross and J.N. Osbon. 2006. Reflections on investigating emotion in educational activity settings. Educational Psychology Review, 18: 343–360.

Scott, E.C. 2009. Evolution vs. Creationism: An Introduction. Westport, CT, Greenwood Press.

Scott, E.C. 2014. Equipping Scientists to Better Understand and Converse with Religious Communities. Paper presented at the AAAS Annual Meeting—Meeting Global Challenges: Discovery and innovation, Chicago.

Shapiro, S. 2009. Revisiting the teachers' lounge: Reflections on emotional experience and teacher identity. Teaching and Teacher Education, 4: 1–6.

Sheinman, N. and L.L. Hadar (Eds.). 2016. Mindfulness in education as a whole of school approach: Principles, insights and outcomes. pp. 77–101. *In*: Ditrich, T., R. Willis and B. Lovegrove (eds.). Mindfulness and Education: Research and Practice. Newcastle, NSW: Cambridge Scholars Publishing.

Shermer, M. 2006. Why Darwin Matters: The Case Against Intelligent Design. New York, Times Books.

Shubin, N. 2009. Your Inner Fish: A Journey into the 3.5-Billion-Year History of the Human Body. New York, Random House, Inc.

Siegel, D. (Producer). 2015a. Myths of the Adolescent Brain - Dan Siegel. Retrieved from https://www.youtube.com/watch?v=0A9LcNt06N8&pbjreload=10.

Siegel, D.J. 2015b. The Developing Mind: How Relationships and the Brain Interact to Shape Who We Are (Second Edition), The Guilford Press.

Siegel, R.D. 2014. The Science of Mindfulness: A Research-Based Path to Well-Being. Chantilly, Virginia, The Teaching Company.

Silberman, I. 2005. Religious violence, terrorism, and peace: A meaning-system analysis. pp. 529–549. *In*: Paloutzian, R.F. and C.L. Park (Eds.). Handbook of the Psychology of Religion and Spirituality. New York, NY, US: Guilford Press.

Simonti, C.N., B. Vernot, L. Bastarache, E. Bottinger, D.S. Carrell, R.L. Chisholm et al. 2016. The phenotypic legacy of admixture between modern humans and Neandertals. Science, 351: 737–741. doi:10.1126/science.aad2149.

Sinatra, G.M., S.H. Broughton and D. Lombardi. 2014. Emotions in science education. pp. 415–436. *In*: Pekrun, R. and L. Linnenbrink-Garcia (Eds.). International Handbook of Emotions in Education. London: Routledge.

Solomon, S., J. Greenberg and T. Pyszczynski. 2015. The Worm at the Core: On the Role of Death in Life, Random House.

Stammer, L.B. 1996. Billy Graham Program Takes Cue From MTV. LA Times.

Steadman, L.B. and C.T. Palmer. 2008. Supernatural and Natural Selection. New York, Routledge.

Strawbridge, W.J., R.D. Cohen, S.J. Shema and G.A. Kaplan. 1997. Frequent attendance at religious services and mortality over 28 years. American Journal of Public Health, 87: 957–961.

Sutton, R.E. and K.F. Wheatley. 2003. Teachers' emotions and teaching: A review of the literature and direction for future research. Educational Psychology Review, 15: 327–358.

Swim, J.K. and J. Fraser. 2013. Fostering hope in climate change educators. Journal of Museum Education, 38: 286–297. doi:10.1179/1059865013z.00000000031.

Symmes, P. 2010. Turkey: Archeological dig reshaping human history. Newsweek.

Sznycer, D., J. Tooby, L. Cosmides, R. Porat, S. Shalvi and E. Halperin. 2016. Shame closely tracks the threat of devaluation by others, even across cultures. Proceedings of the National Academy of Sciences of the United States of America, 113: 2625–2630. doi:10.1073/pnas.1514699113.

Tabak, N.T., W.P. Horan and M.F. Green. 2015. Mindfulness in schizophrenia: Associations with self-reported motivation, emotion regulation, dysfunctional attitudes, and negative symptoms. Schizophr Res., 168: 537–542. doi:10.1016/j.schres.2015.07.030.

Taber, K.S. 2017. Beliefs and science education. pp. 53–67. *In*: Taber, K.S. and B. Akpan (Eds.). Science Education: Sense Publishers. Rotterdam/Boston/Taipei.

Taddonio, P. 2018. How Billy Graham Helped Merge Patriotism and Christianity. Frontline. Retrieved from https://www.pbs.org/wgbh/frontline/article/how-billy-graham-helped-merge-patriotism-and-christianity/.

Tang, Y.-Y., B.K. Hölzel and M.I. Posner. 2015. The neuroscience of mindfulness meditation. Nature Reviews Neuroscience, 16: 213. doi:10.1038/nrn3916.

Taren, A.A., J.D. Creswell and P.J. Gianaros. 2013. Dispositional Mindfulness Co-Varies with Smaller Amygdala and Caudate Volumes in Community Adults. PLOS ONE, 8: e64574. doi:10.1371/journal.pone.0064574.

Tayler, J. 2015. Can Religion and Science Coexist? The Atlantic. Retrieved from https://www.theatlantic.com/politics/archive/2015/07/religion-science-coexist-faith-versus-fact-coyne/396362/?utm_source=atlfb.

Teffer, K. and K. Semendeferi. 2012. Human prefrontal cortex: Evolution, development, and pathology. pp. 191–218. *In*: Hofman, M.A. and D. Falk (Eds.). Evolution of the Primate Brain: From Neuron to Behavior (Vol. 195): Elsevier.

Think:Kids. 2014. Regulate, Relate, Reason. Think:Kids Rethinking Challenging Kids. Retrieved from http://www.thinkkids.org/regulate-relate-reason/.

Tobin, K. 2012. Sociocultural perspectives on science education. pp. 3–17. *In*: Fraser, B.J., C.J. McRobbie and K.G. Tobin (Eds.). Second International Handbook of Science Education. Dordrecht: Springer.

Tobin, K., K. Alexakos and M. Powietrzynska. 2015. Mindfully speaking & mindfully listening heuristics (white paper). New York, The City University of New York.

Utter, G. and J. True. 2004. Conservative Christians and Political Participation—A Reference Handbook. Santa Barbara, California, ABC Clio.

Wacker, G. 2000. The Rise of Fundamentalism. TeacherServe. Retrieved from http://nationalhumanitiescenter.org/tserve/twenty/tkeyinfo/fundam.htm.

Wacker, G. 2014. America's Pastor: Billy Graham and the Shaping of a Nation, Harvard University Press.

Wallis, C. 2005. The Evolution Wars. TIME, pp. 52–59.

Wang, S.Z., S. Li, X.Y. Xu, G.P. Lin, L. Shao, Y. Zhao et al. 2010. Effect of slow abdominal breathing combined with biofeedback on blood pressure and heart rate variability in prehypertension. Journal of alternative and complementary medicine (New York, N.Y.), 16: 1039–1045. doi:10.1089/acm.2009.0577.

Weare, K. 2012. Evidence for the Impact of Mindfulness on Children and Young People.

Weinberg, S. 1984. The First Three Minutes Mass Market.

Wexler, J. 2010. From the Classroom to the Courtroom: Intelligent Design and the Constitution. Evo. Edu. Outreach, 3: 215–224.

Whitbourne, S.K. 2010. In-groups, out-groups, and the psychology of crowds: Does the ingroup-outgroup bias form the basis of extremism? Retrieved from https://www.psychologytoday.com/intl/blog/fulfillment-any-age/201012/in-groups-out-groups-and-the-psychology-crowds.

Wolf, G. 2006. The Church of the Non-Believers. Retrieved from https://www.wired.com/2006/11/atheism/?pg=1&topic=atheism&topic_set=.

Worrall, S. 2015. In Age of Science, Is Religion 'Harmful Superstition'? National Geographic. Retrieved from https://news.nationalgeographic.com/2015/05/150531-religion-science-faith-healing-atheism-people-ngbooktalk/.

Yurgelun-Todd, D. 2002. Inside the Teenage Brain.

Zajonc, A. 2016. Contemplation in Education Handbook of Mindfulness in Education (pp. 17–28). New York: Springer.

Zelazo, P.D. and W. Cunningham. 2007. Executive function: Mechanisms underlying emotion regulation pp. 135–158. *In*: Gross, J. (Ed.). Handbook of Emotion Regulation. New York: Guilford Press.

Zembylas, M. 2002. Constructing genealogies of teachers' emotions in science teaching. Journal of Research in Science Teaching, 39: 79–103.

Zembylas, M. 2004. Young children's emotional practices while engaged in long-term science investigations. Journal of Research in Science Education, 41: 213–238.

Zembylas, M. 2005. Three Perspectives on Linking the Cognitive and the Emtional in Science Learning: Conceptual Change, Socio-Constructivism and Poststructuralism. Studies in Science Education, 41: 91–116.

Zenner, C., S. Herrnleben-Kurz and H. Walach. 2014. Mindfulness-based Interventions in Schools—A Systematic Review and Meta-analysis. Frontiers in Psychology, 5: 603.

Zoogman, S., S.B. Goldberg, W.T. Hoyt and L. Miller. 2015. Mindfulness Interventions with Youth: A Meta-analysis. Mindfulness, 6: 290–302.

# Index

# Color Plate Section

## Chapter 6

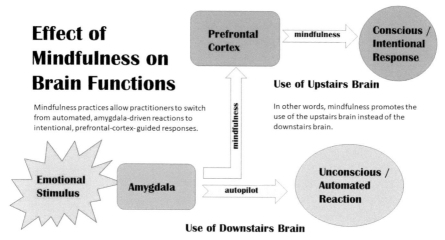

**Figure 21:** Illustration of how mindfulness affects how the brain functions, allowing a movement from automatic, emotionally-driven reactions to conscious, intentional responses.

## Chapter 7

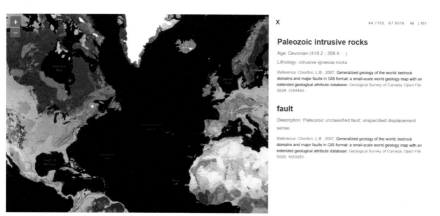

**Figure 25:** Example of an interactive geologic map where students could find potential dig sites for Devonian fossil search.

## Afterword

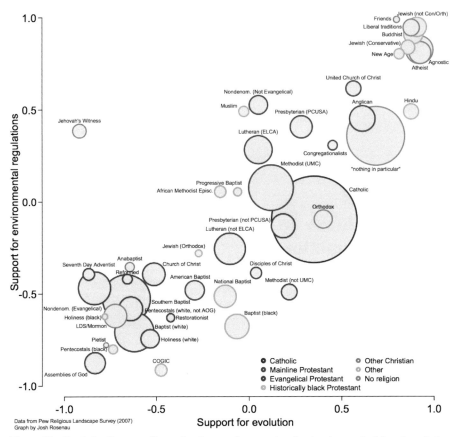

**Figure 28:** Graph by Rosenau illustrating how anthropogenic climate change denial and evolution opposition coincide, where as acceptance of evolutionary theory increases, so does support for environmental regulations.